The IEE

Code of Practice for In-service Inspection and Testing of Electrical Equipment

Published by: The Institution of Electrical Engineers
Savoy Place, LONDON
United Kingdom. WC2R 0BL

First published 1994
2nd edition 2001
Reprinted 2003, with new cover

Copies may be obtained from:
The IEE
PO Box 96, STEVENAGE
United Kingdom. SG1 2SD.

Tel: 01438 767 328
Fax: 01438 742 792
Email: sales@iee.org
http://www.iee.org/Publish/Books/WireAssoc/

ISBN 0 85296 776 4

CODE OF PRACTICE
In-service Inspection and Testing of Electrical Equipment

Contents

Co-operating Organisations

The Institution of Electrical Engineers acknowledges the contribution made by the following representatives of organisations in the preparation of this Code of Practice.

S MacConnacher BSc
Association of Manufacturers of Domestic Electrical Appliances

A J A Hinsley BSc CEng MIMgt
BETNET

P D Stokes MA, CEng, MRAeS
British Electrotechnical Approvals Board

G Scott BEng ACGI MSc CEng MIEE
Department of Trade and Industry

H R Lovegrove IEng FIIE
Electrical Contractors' Association

D Millar IEng MIIE MILE
SELECT (Electrical Contractors' Association of Scotland)

S A Johnson CEng MIEE
Electricity Association

F W Pearson CEng MIEE
Federation of Electronic Industries

R Piper
GAMBICA Association Ltd

R T R Pilling CEng MIEE
Health and Safety Executive

K Minton
Hire Association Europe

W H Wright CEng MIEE
Institution of Electrical Engineers

R F Armstrong IEng MIEE
Institution of Incorporated Engineers

J Ware BSc CEng MIEE
National Inspection Council for Electrical Installation Contracting

A Smyth MSc(Eng) DIC
Portable Electric Tool Manufacturers' Association

SUMMARY

This Code of Practice provides guidance on the inspection and testing of electrical appliances, luminaires, information technology equipment supplied by a plug and socket and similar equipment. It is written both for administrators with responsibility for electrical maintenance (who may have little technical knowledge) and the staff who carry out the inspections and tests.

Inspection and testing is reduced to three activities:

(1) user checks

(2) formal visual inspections (without tests)

(3) combined inspections and tests.

Part 1 provides guidance on what work needs to be done, and whether it can be carried out in-house. Advice is included on the law, training, procedures and documentation.

Part 2 is written for those carrying out the practical work and details the inspections and tests.

Part 3 comprises six appendices containing information and guidance, e.g. model forms.

PART 1 ADMINISTRATION OF IN-SERVICE INSPECTION AND TESTING

1. SCOPE

This Code of Practice has been prepared with a view to meeting the needs of:

(a) those with administrative responsibilities, either within or contracted to an organisation or authority, for maintenance of electrical equipment - see Part 1

(b) those undertaking practical inspection and testing of electrical equipment, who are recommended to read the whole document. Part 2 (in pink) identifies test data they will need to refer to frequently.

The equipment within the scope of this document includes electrical appliances for household and similar use, certain IT equipment supplied by plug and socket, luminaires, and similar equipment.

The document considers the obligations imposed on persons by the Health and Safety at Work etc. Act 1974, the requirements of the Provision and Use of Work Equipment Regulations 1998, the Management of Health and Safety at Work Regulations 1999, the Workplace (Health, Safety and Welfare) Regulations 1992 and the Electricity at Work Regulations 1989. It is intended to give administrators:

sufficient knowledge and understanding so that they may make decisions concerning arrangements for electrical equipment testing, including the frequency of inspections and tests

some understanding of the service they should receive from any organisation carrying out inspection and testing, including documents, and inspection and testing frequencies

assistance in deciding as to whether the work could be carried out in-house.

The document should also help those persons who have specific responsibility for carrying out the electrical inspections and tests, either in-house or on a contractual basis for a client. The Code of Practice provides specific information on the inspections and tests necessary.

The guidance is limited to non-specialised situations such as offices, shops, hotels, schools and general industrial locations. Where premises have mixed use, the most appropriate mix of frequencies of inspection and of testing will need to be adopted.

The guidance does not encompass the legislation relating to supply of equipment, whether new or second-hand, to a third party by way of sale, hire or other method.

This advice is mainly concerned with electrical inspection and testing as part of the maintenance of electrical equipment. However, it is important that the electrical installation to which the equipment is connected is not faulty and brief reference is made to fixed installation testing. Further guidance on inspection and testing of the electrical installation can be obtained from the Institution of Electrical Engineers in Guidance Note 3, Inspection & Testing.

The types of equipment covered by this Code of Practice are described in paragraph 5.1.

2. DEFINITIONS

To assist the reader the following definitions are provided:

BASIC INSULATION Insulation applied to live parts to provide basic protection against electric shock and which does not necessarily include insulation used exclusively for functional purposes.

CLASS I EQUIPMENT (see BS 2754). Equipment in which protection against electric shock does not rely on basic insulation only, but which includes means for the connection of exposed-conductive-parts to a protective conductor in the fixed wiring of the installation.

CLASS II EQUIPMENT (see BS 2754). Equipment in which protection against electric shock does not rely on basic insulation only, but in which additional safety precautions such as supplementary insulation are provided, there being no provision for the connection of exposed metalwork of the equipment to a protective conductor and no reliance upon precautions to be taken in the fixed wiring of the installation.

Notes

1 Such an appliance may be one of the following types:
   a) an appliance having a durable and substantially continuous enclosure of insulating material which envelops all metal parts, with the exception of small parts, such as nameplates, screws and rivets, which are isolated from live parts by insulation at least equivalent to reinforced insulation; such an appliance is called an insulation-encased Class II appliance;
   b) an appliance having a substantially continuous metal enclosure, in which double insulation or reinforced insulation is used throughout; such an appliance is called a metal-encased Class II appliance;
   c) an appliance which is a combination of types a) and b).

2 The enclosure of an insulation-encased Class II appliance may form part or the whole of the supplementary insulation or of the reinforced insulation.

3 If an appliance with double insulation or reinforced insulation throughout has provision for earthing, it is considered to be a Class I or OI appliance.

4 Class II appliances may incorporate means for maintaining the continuity of protective circuits, provided that such means are within the appliance and are insulated from conductive accessible parts by supplementary insulation.

CLASS III EQUIPMENT (see BS 2754). Equipment in which protection against electric shock relies on the supply from a separated extra-low voltage source (SELV), such as an isolating transformer to BS EN 61558.

CORD SET An assembly consisting of a detachable flexible cable or cord fitted with a plug and a connector, intended for the connection of electrical equipment to the electrical supply.

DANGER Risk of injury to persons (and livestock where expected to be present) from:

(i) fire, electric shock and burns arising from the use of electrical energy, and

(ii) mechanical movement of electrically controlled equipment, in so far as such danger is intended to be prevented by electrical emergency switching or by electrical switching for mechanical maintenance of non-electrical parts of such equipment.

DOUBLE INSULATION Insulation comprising both basic insulation and supplementary insulation.

EARTHING Connection of the exposed-conductive-parts of electrical equipment to the main earthing terminal of an electrical installation.

ELECTRIC SHOCK A dangerous physiological effect resulting from the passing of an electric current through a human body or livestock.

EMERGENCY SWITCHING Rapid cutting off of electrical energy to remove any unexpected hazard to persons, livestock or property.

ENCLOSURE A part providing an appropriate degree of protection of equipment against certain external influences and a defined degree of protection against contact with live parts from any direction.

EXPOSED-CONDUCTIVE-PART A conductive part of equipment which can be touched and which is not a live part but which may become live under fault conditions.

FLEXIBLE CABLE Cable whose structure and materials make it suitable to be flexed while in service.

FLEXIBLE CORD A flexible cable in which the cross-sectional area of each conductor does not exceed 4 mm^2.

INSULATION Suitable non-conductive material enclosing, surrounding or supporting a conductor.

LIVE PART A conductor or conductive part intended to be energised in normal use, including a neutral conductor.

NEUTRAL CONDUCTOR A conductor connected to the neutral point of a system and contributing to the transmission of electrical energy.

PAT INSTRUMENT Portable appliance test instrument.

PHASE CONDUCTOR A conductor of an a.c. system for the transmission of electrical energy other than a neutral conductor or a protective conductor.

PROTECTIVE CONDUCTOR CURRENT Electric current which flows in a protective conductor under normal operating conditions.

PROTECTIVE EARTHING CONDUCTOR A conductor used for some measures of protection against electric shock and intended for connecting together the exposed-conductive-parts of electrical equipment to the main earth terminal of an electrical installation.

REINFORCED INSULATION Single insulation applied to live parts, which provides a degree of protection against electric shock equivalent to double insulation under the conditions specified in the relevant standard. The term 'single insulation' does not imply that the insulation must be one homogeneous piece. It may comprise several layers which cannot be tested singly as supplementary or basic insulation.

SUPPLEMENTARY INSULATION Independent insulation applied in addition to basic insulation in order to provide protection against electric shock in the event of a failure of basic insulation.

TOUCH CURRENT Electric current through a human body or through an animal's body when it touches one or more accessible parts of an installation or equipment.

3. THE LAW

3.1 General

This Code of Practice has been prepared by the Institution of Electrical Engineers with a view to determining the inspections and tests necessary to ensure that electrical equipment is maintained properly so as to prevent danger. Although reference is made to legislation, this chapter should not be considered as legal advice. In cases of doubt, the specific legislation mentioned should be consulted and legal advice obtained.

In recent years the responsibilities for safety of persons at work have been prescribed in much legislation, some of which is listed in Appendix II.

The legislation of specific relevance to electrical maintenance is the Health and Safety at Work etc. Act 1974, the Management of Health and Safety at Work Regulations 1999, the Electricity at Work Regulations 1989, the Workplace (Health, Safety and Welfare) Regulations 1992 and the Provision and Use of Work Equipment Regulations 1998.

The Health and Safety at Work etc. Act 1974 puts a duty of care upon both employer (sections 2, 3 and 4 etc) and employee (section 7) to ensure the safety of **all** persons using the work premises. This includes the self-employed.

The Management of Health and Safety at Work Regulations 1999 state:

"Every employer shall make a suitable and sufficient assessment of:

(a) the risks to the health and safety of his employees to which they are exposed whilst they are at work, and

(b) the risks to the health and safety of persons not in his employment arising out of or in connection with the conduct by him of his undertaking". (Regulation 3(1))

The Provision and Use of Work Equipment Regulations 1998 state:

"Every employer shall ensure that work equipment is so constructed or adapted as to be suitable for the purpose for which it is used or provided". (Regulation 4(1))

The Provision and Use of Work Equipment Regulations 1998 (PUWER) cover most risks that can result from using work equipment. With respect to risks from electricity, compliance with the Electricity at Work Regulations 1989 is likely to achieve compliance with PUWER regulations 5-9, 19 and 22.

PUWER only applies to work equipment used by workers at work. This includes all work equipment (fixed, portable or transportable) connected to a source of electrical energy. PUWER does not apply to the fixed installations in a building. The electrical safety of these installations is dealt with only by the Electricity at Work Regulations.

The Electricity at Work Regulations 1989 state:

"As may be necessary to prevent danger, all systems shall be maintained so as to prevent, so far as is reasonably practicable, such danger". (Regulation 4(2))

"'System' means an electrical system in which all the electrical equipment is, or may be, electrically connected to a common source of electrical energy and includes such source and such equipment". (Regulation 2(1))

"Electrical equipment includes anything used, intended to be used or installed for use, to generate, provide, transmit, transform, rectify, convert, conduct, distribute, control, store, measure or use electrical energy". (Regulation 2(1))

3.2 Scope of the legislation

It is clear that the combination of the Health and Safety at Work etc. Act 1974, the Provision and Use of Work Equipment Regulations 1998 and the Electricity at Work Regulations 1989 apply to all electrical equipment used in, or associated with, places of work. The scope extends from distribution systems, be they 400 kV or simply those in buildings, down to the smallest piece of electrical equipment such as a hairdryer, a VDU, a telephone or even in some situations battery-operated equipment.

3.3 Who is responsible?

Everyone at work has their responsibilities including, in certain circumstances, trainees. However, because of the all-embracing responsibilities of all persons this does not minimise the duties of particular persons. Regulation 3 of the Electricity at Work Regulations recognises a responsibility (control) that employers and many employees have for electrical systems.

"It shall be the duty of every employer and self-employed person to comply with the provisions of these Regulations in so far as they relate to matters which are within his control.

It shall be the duty of every employee while at work:

(a) to co-operate with his employer so far as is necessary to enable any duty placed on that employer by the provisions of these Regulations to be complied with; and

(b) to comply with the provisions of these Regulations in so far as they relate to matters which are within his control".

The Provision and Use of Work Equipment Regulations 1998 requires every employer to ensure that equipment is suitable for the use for which it is provided (Reg 4(1)) and only used for work for which it is suitable (Reg 4(3)). They require every employer to ensure equipment is maintained in good order (Reg 5) and inspected as necessary to ensure it is maintained in a safe condition (Reg 6).

This Code of Practice considers normal business premises such as shops, offices and workplaces and restricts advice to non-specialist installations and equipment that are commonly encountered.

3.4 Maintenance

Regulation 4(2) of the Electricity at Work Regulations 1989 states:

"As may be necessary to prevent danger, all systems shall be maintained so as to prevent, so far as is reasonably practicable, such danger."

Regulation 5 of the Provision and Use of Work Equipment Regulations 1998 states:

"Every employer shall ensure that work equipment is maintained in an efficient state, in efficient working order and in good repair".

The Approved Code of Practice and Guidance document to the Provision and Use of Work Equipment Regulations 1998 (L22) states that 'efficient' relates to how the condition of the equipment might affect health and safety; it is not concerned with productivity.

3.5 Inspection

The Provision and Use of Work Equipment Regulations 1998 include a specific requirement that where the safety of work equipment depends on installation conditions, and where conditions of work are liable to lead to deterioration, the equipment shall be inspected (Reg 6).

4. FIXED ELECTRICAL INSTALLATIONS

The Provision and Use of Work Equipment Regulations 1998 require that where safety of work equipment depends on installation conditions, the installation is inspected on completion, and if subject to deterioration inspected again at suitable intervals. Records are to be kept of such inspections (Reg 6).

The safety and proper functioning of certain portable appliances and equipment depend on the integrity of the fixed installation. It is important to establish a system for the inspection and testing of fixed installations as well as for the inspection and testing of portable appliances and equipment. Requirements for the inspection and testing of fixed installations are given in BS 7671 and guidance is provided in IEE Guidance Note 3, Inspection & Testing. Guidance Note 3 provides information on the necessary competence of those carrying out the tests and gives advice on the frequency of inspection.

Attention is drawn to the existence of certain specialised fixed electrical installations such as petrol filling stations, locations with combustible dusts and locations with potentially explosive atmospheres, in which only suitable portable appliances and equipment are permitted. See Appendix I for information on relevant standards and Appendix III for HSE and HSC Guidance Notes.

5. ELECTRICAL EQUIPMENT

5.1 Equipment types

The following types of electrical equipment are covered by this Code of Practice:

Portable appliance

An appliance of less than 18 kg in mass that is intended to be moved while in operation or an appliance which can easily be moved from one place to another, e.g. toaster, food mixer, vacuum cleaner, fan heater.

Movable equipment
(sometimes called transportable)

This is equipment which is either:

- 18 kg or less in mass and not fixed, e.g. electric fire, or
- equipment with wheels, castors or other means to facilitate movement by the operator as required to perform its intended use, e.g. air conditioning unit.

Hand-held appliances or equipment

This is portable equipment intended to be held in the hand during normal use, e.g. hair dryer, drill, soldering iron.

Stationary equipment or appliances

This equipment has a mass exceeding 18 kg and is not provided with a carrying handle, e.g. refrigerator, washing machine.

Fixed equipment/appliances

This is equipment or an appliance which is fastened to a support or otherwise secured in a specified location, e.g. bathroom heater, towel rail.

Appliances/equipment for building in

This equipment is intended to be installed in a prepared recess such as a cupboard or similar. In general, equipment for building in does not have an enclosure on all sides because on one or more of the sides, additional protection against electric shock is provided by the surroundings e.g. a built-in electric cooker.

Information technology equipment (business equipment)

Information technology equipment includes electrical business equipment such as computers and mains powered telecommunications equipment, and other equipment for general business use, such as mail processing machines, electric plotters, trimmers, VDUs, data terminal equipment, typewriters, telephones, printers, photo-copiers, power packs.

Extension leads

The use of extension leads should be avoided where possible. If used, they should be tested as portable appliances. It is recommended that 3-core leads (including a protective earthing conductor) be used.

A standard 13 A 3-pin extension socket-outlet with a 2-core cable should never be used even if the appliance to be used is Class II, as it would not provide protection against electric shock if used at any time with an item of Class I equipment.

The length of an extension lead for general use should not exceed the following:

core area	maximum length
1.25 mm^2	12 metres
1.5 mm^2	15 metres
2.5 mm^2	25 metres*

* 2.5 mm^2 cables are too large for standard 13 A plugs, but they may be used with BS EN 60309 industrial plugs.

These maximum lengths are not applicable to the flex of an appliance, for guidance refer to paragraph 15.13.

If extension lead lengths do exceed the above, they shall be protected by a 30 mA RCD manufactured to BS 7071.

5.2 British and European standards

To encourage free trade within the European Union, existing national standards are being harmonized and converted to European standards. Compliance with these standards gives assurance to purchasers that appliances and equipment have been designed and constructed to a standard which will ensure that in normal use they function safely and without danger.

In order to check compliance, manufacturers have to perform a series of tests on the appliance and its components as required by the standard. The appliance must pass these tests if it is to be said that the appliance complies with the standard. A list of some of the safety standards for electrical equipment is given in Appendix I. The tests detailed in these standards are not suitable for in-service testing.

This Code recommends in-service inspections and tests which can be applied generally to equipment and appliances in normal use.

Routine manufacturers' tests are not required for general in-service testing, but may be applied to new appliances or after repair. Further information is provided in Appendix IV 5.3.

6. THE ELECTRICAL TESTS

6.1 General

It is emphasised that care must be taken when testing electrical equipment to avoid damaging otherwise satisfactory equipment by the application of test voltages and currents.

Four test situations are recognised:

(a) Type testing to an appropriate standard e.g. BS EN 60835

(b) Production testing

(c) In-service testing (covered by this Code of Practice)

(d) Testing after repair

6.2 Manufacturers' tests

6.2.1 Type testing

Type testing is carried out by test houses or manufacturers to assess compliance with a standard (British or European). The tests are usually destructive, making the appliance unsuitable for sale or use.

6.2.2 Production testing

Production testing is carried out by manufacturers to ensure that manufactured appliances are in accordance with the appropriate standard.

These tests are designed to be applied to new appliances uncontaminated by dust or lubricants. It may be appropriate to apply them to equipment in new or as-new condition following refurbishment or repair. For this reason and for general guidance they are included in Appendix IV.

6.3 In-service testing

This is the testing carried out as a routine to determine whether the equipment is in satisfactory condition. It is the testing that most users of this Code of Practice will require to be carried out. It is not as onerous as production line testing and testers must always be aware of its limitations.

In-service testing will involve the following:

(a) preliminary inspection

(b) earth continuity tests (for Class I equipment)

(c) insulation testing (which may sometimes be substituted by earth leakage measurement)

(d) functional checks.

However, inspection of equipment is of particular importance. There is no substitute for a properly carried out visual inspection. There are many faults which can be seen during inspection that will not necessarily be identified by electrical tests, e.g. cracked cases, loose connections, damaged flexes, signs of overheating.

6.4 Testing after repair

A repairer may wish to subject an appliance either to the production tests paragraph 6.2.2 above, or the in-service tests paragraph 6.3. The repairer will need to make a decision based on the condition of the appliance and the nature of the repairs he has carried out. The repairer must be knowledgeable about the equipment and able to make the decision. Guidance on testing after repair may be obtained from the manufacturer.

7. IN-SERVICE INSPECTION AND TESTING

7.1 Inspection

In-service inspection of equipment is essential to ensure safety. This can often be carried out by the user of the equipment, and in some circumstances (e.g. low risk environment with Class II equipment) may be all that is necessary. Inspection should always precede testing. This Code of Practice recommends a regime of time and cost effective inspections and tests.

7.2 Categories of inspection and testing

Three categories of in-service inspection and testing are referred to in this recommendation:

(a) user checks; faults are to be reported and logged, but no record is required if no fault is found

(b) formal visual inspections; inspections without tests the results of which, satisfactory or unsatisfactory, are recorded

(c) combined inspections and tests, the results of which are recorded.

Details of these inspections and tests are provided in Part 2.

7.3 Frequency of inspection and testing

The relevant requirement of the Electricity at Work Regulations 1989 is that equipment shall be maintained so as to prevent danger. Inspection and testing are means of determining whether maintenance is required. The frequency of inspection and testing will depend upon the likelihood of maintenance being required and the consequence of lack of maintenance. No rigid guidelines can be laid down, but factors influencing the decision will include the following:

(a) the environment - equipment installed in a benign environment will suffer less damage than equipment in an arduous environment

(b) the users - if the users of equipment report damage as and when it becomes evident, hazards will be avoided. Conversely, if equipment is likely to receive unreported abuse, more frequent inspection and testing is required

(c) the equipment construction - the safety of a Class I appliance is dependent upon a connection with the earth of the fixed electrical installation. If the flexible cable is damaged the connection with earth can be lost

The safety of Class II equipment is not dependent upon the fixed electrical installation.

If equipment is known to be Class II, in **a low risk environment** such as an office, recorded testing (but not inspection) may be omitted - see Table 1.

(d) the equipment type - appliances which are hand-held are more likely to be damaged than fixed appliances. If they are also Class I the risk of danger is increased, as safety is dependent upon the continuity of the protective conductor from the plug to the appliance.

Table 1 provides guidance on **INITIAL** frequencies of inspection and testing. However, the frequency must depend upon the factors above, i.e. any circumstance which may affect the continuing safety of the equipment. The frequency of any recurring damage should be noted and related to the frequency of inspection and testing.

It is appropriate to restate that the most important check that can be carried out on a piece of equipment, particularly portable appliances or hand-held tools, is the visual inspection.

If the equipment cannot be disconnected routinely by the user to facilitate a user inspection, this must be taken into account when determining the frequency of recorded inspection.

7.4 Review of frequency of inspection and testing

The intervals between checks, formal inspections and tests must be kept under review, particularly until patterns of failure/damage, if any, are determined.

Particularly close attention must be paid to initial checks, formal inspections and tests to see if there is a need to reduce the intervals or change the equipment or its use. After the first few inspections/tests consideration can be given to increasing the intervals or, reducing them.

TABLE 1 - SUGGESTED INITIAL FREQUENCY OF INSPECTION AND TESTING OF EQUIPMENT

	Type of Premises	Type of Equipment Note (1)	User Checks Note (2)	Class I		Class II Note (4)	
				Formal Visual Inspection Note (3)	Combined Inspection and Testing Note (5)	Formal Visual Inspection Note (3)	Combined Inspection and Testing Note (5)
	1	2	3	4	5	6	7
1	Construction sites	S	None	1 month	3 months	1 month	3 months
		IT	None	1 month	3 months	1 month	3 months
	110 V }	M#	weekly	1 month	3 months	1 month	3 months
	equipment }	P#	weekly	1 month	3 months	1 month	3 months
	}	H#	weekly	1 month	3 months	1 month	3 months
2	Industrial including commercial kitchens	S	weekly	None	12 months	None	12 months
		IT	weekly	None	12 months	None	12 months
		M	before use	1 month	12 months	3 months	12 months
		P	before use	1 month	6 months	3 months	6 months
		H	before use	1 month	6 months	3 months	6 months
3	Equipment used by the public	S	Note (6)+	monthly	12 months	3 months	12 months
		IT	Note (6)+	monthly	12 months	3 months	12 months
		M	Note (6)+	weekly	6 months	1 month	12 months
		P	Note (6)+	weekly	6 months	1 month	12 months
		H	Note (6)+	weekly	6 months	1 month	12 months
4	Schools	S	weekly+	None	12 months	12 months	48 months
		IT	weekly+	None	12 months	12 months	48 months
		M	weekly+	4 months	12 months	4 months	48 months
		P	weekly+	4 months	12 months	4 months	48 months
		H	before use+	4 months	12 months	4 months	48 months
5	Hotels	S	None	24 months	48 months	24 months	None
		IT	None	24 months	48 months	24 months	None
		M	weekly	12 months	24 months	24 months	None
		P	weekly	12 months	24 months	24 months	None
		H	before use	6 months	12 months	6 months	None
6	Offices and shops	S	None	24 months	48 months	24 months	None
		IT	None	24 months	48 months	24 months	None
		M	weekly	12 months	24 months	24 months	None
		P	weekly	12 months	24 months	24 months	None
		H	before use	6 months	12 months	6 months	None

(1) S Stationary equipment
IT Information technology equipment
M Movable equipment
P Portable equipment
H Hand-held equipment

(2) User checks are not recorded unless a fault is found.

(3) The formal visual inspection may form part of the combined inspection and tests when they coincide, and must be recorded see 7.2b.

(4) If class of equipment is not known, it must be tested as Class I.

(5) The results of combined inspections and tests are recorded see 7.2c.

(6) For some equipment such as children's rides a daily check may be necessary

(+) By supervisor/teacher/member of staff

110 V earthed centre tapped supply. 230 V portable or hand-held equipment must be supplied via a 30 mA RCD and inspections and tests carried out more frequently.

The information on suggested initial frequencies given above is more detailed and specific than HSE guidance, but is not considered to be inconsistent with it.

8. PROCEDURES FOR IN-SERVICE INSPECTION AND TESTING

8.1 Procedures

8.1.1 The basic requirement

There is no specific requirement in legislation concerned with in-service maintenance to inspect or test equipment but the Electricity at Work Regulations 1989 Regulation 4(2) requires that:

"As may be necessary to prevent danger, all systems shall be maintained so as to prevent, so far as is reasonably practicable, such danger".

Inspection and testing alone will not result in compliance with the law.

The requirement is to maintain electrical equipment as necessary to prevent danger. This means that the requirements of the Electricity at Work Regulations are not met simply by carrying out tests on equipment - the aim is to ensure that equipment is maintained in a safe condition.

8.1.2 Fixed installations

The procedures described here are for the inspection and testing of electrical equipment. Similar procedures must be followed for the fixed electrical installation.

Those testing equipment must advise the responsible person of his/her responsibilities with respect to the fixed installation and, if the instructions do not include the inspection and testing of socket-outlets, the client must be advised in writing of the need to ensure, by periodic testing (see IEE Guidance Note 3), that the fixed installation is satisfactory. The reliance of Class I appliances on a good connection with earth for safety must be emphasised.

8.1.3 Equipment

The following procedure is necessary:

(a) inspection and testing as necessary of all electrical equipment including:

- fixed equipment
- portable equipment
- electrical tools
- electrical instruments etc.

so that necessary maintenance and repair is identified.

(b) following inspection and testing, necessary maintenance and repairs must be carried out or the equipment removed from use.

Adequate records of all maintenance activities must be kept including:

- inspections
- testing
- repairs
- procurement of equipment.

Organisations whose equipment is being inspected and tested will need to assist in identifying the tests to be carried out. Manufacturers' recommendations and past test results must be requested by the tester from the responsible person. This is particularly appropriate to business and telecommunications equipment, where it may be necessary to discuss the appropriate tests with the supplier of the equipment. Some tests, particularly insulation resistance tests, may damage such equipment. If there is doubt, a minimum test should be carried out, and the customer advised that, after discussion with the supplier and down-loading of data, it may be necessary to carry out further tests.

IF IN DOUBT DO NOT CARRY OUT THE INSULATION RESISTANCE TEST

8.2 Documentation

The Provision and Use of Work Equipment Regulations 1998 contain no specific requirement to keep maintenance records, but the Health and Safety Commission recommend a maintenance record for high risk equipment.

Although there is no requirement in the Electricity at Work Regulations 1989 to keep records of equipment and of inspections and tests, the HSE Memorandum of Guidance on these Regulations advises that records of maintenance including tests should be kept throughout the working life of equipment. These records are a useful management tool for reviewing the frequency of inspection and testing, and without such records duty holders cannot be certain that the inspection and testing has actually been carried out.

The following records (see Appendix V) should be established and maintained:

(a) a register of all equipment (Form Va)

(b) a record of formal and combined inspections and tests (Form Vb)

(c) a repair register (Form Vd)

(d) a register of all faulty equipment (Form Ve)

(e) all equipment formally inspected and tested should be labelled as per paragraph 8.3 (Form Vc).

These records may be retained on paper or on electronic memory provided that reasonable precautions are taken with respect to security. **Previous test results must be made available to subsequent testers.**

The following records should be maintained by the agency carrying out the inspection and testing.

(a) copy of the formal visual inspection and combined inspection and test results

(b) register of all equipment repaired.

These may be held as paper or electronic records.

8.3 Labelling

All equipment which requires routine inspection and/or testing must be clearly identifiable. This is usually achieved by labelling of the equipment. The information provided should consist of an identification code to enable the equipment to be uniquely identifiable even if several similar items exist within the same premises. An indication of the current safety status of the equipment must also be included (e.g. whether the item has PASSed or FAILed the appropriate safety inspection/test). The date on which re-testing is due or the last test date and re-test period should also be stated.

The provision of the above information will not only ease location of the equipment at the time of re-test but will enable non-technical users to become aware of any equipment which is due for re-test, or which should not be used.

Additional information may also be included such as the company name or logo type.

Labels may either be pre-printed and filled in by hand or be machine readable (e.g. bar-coded). The latter is particularly suitable for the identification code. Many PAT testers can read bar-coded labels to set up the instrument to conduct only the tests appropriate for the equipment. Pass/fail information and date should be in text format so that it is readable by others.

Labels may take many forms but should be such that they can be reliably applied to a variety of surfaces. They should be durable and capable of surviving the period between re-tests without undue degradation. In industrial environments the demands on the label are high since it may be subject to contact with oils, solvents, moisture and abrasion. On larger equipment the label should be fixed in a prominent position where it will be clearly seen.

In order to keep proper records, items such as extension leads which may not have serial numbers, will have to be identified with a unique reference number or code fixed to or marked on the equipment.

8.4 Damaged/faulty equipment

If equipment is found to be damaged or faulty on inspection or test, an assessment must be made by a responsible person as to the suitability of the equipment for the use in that particular location. Equipment found to be unsafe must be immediately removed from use. More frequent inspections and tests will not prevent damage to equipment arising from its unsuitability for the use, location or environment. If it is unsuitable it must be replaced by suitable equipment.

8.5 User responsibilities

Staff and users of equipment must be advised that it is their legal responsibility to comply with the Health and Safety at Work etc. Act and the Electricity at Work Regulations by assisting in the maintenance of equipment.

Equipment must be regularly inspected.

Such inspections should initially be carried out at the frequencies indicated in Table 1.

Faulty equipment must NOT be used

Faulty equipment must be labelled, reported and withdrawn from service.

8.6 Provision of test results

Previous inspection and test results of equipment should be provided to persons carrying out in-service inspection and testing.

9. TRAINING

9.1 General

All users of equipment must be encouraged and trained to report formally any faults found or suspected in electrical equipment and appliances.

Faulty equipment must NOT be used

Faulty equipment must be labelled, reported and, if unsafe, removed from use without delay

A system for logging reports of faults on electrical appliances and equipment is necessary, particularly in hotels or offices, where the number of reports can be expected to be numerous. A sample Form Ve can be found in Appendix V.

9.2 Equipment inspection and testing

As outlined above, all users should visually inspect electrical equipment for defects before it is switched on or used.

Managers of premises have a legal responsibility to ensure that the electrical equipment in their charge is safe. To achieve this level of safety through use of this Code of Practice requires three categories of specifically trained person (in addition to a general awareness by all staff of the dangers of electricity) as follows:

(a) managers of the premises

(b) managers of the inspection and testing organisations

(c) those who inspect and test.

The above division of tasks and responsibilities does not presume that one person may not carry out two or all of the required functions. It is, however, stressed that a person must be trained in each of the areas and be competent to undertake the work and interpret the results as appropriate.

To set a standard of training in the UK, the IEE has combined with the City and Guilds of London Institute to provide two courses and associated certificated examinations for

(a) managers of premises and of inspection and testing organisations

and

(b) those who inspect and test.

9.3 Managers

Managers of premises and of inspection and testing organisations are required to know their legal responsibilities as laid down in the Electricity at Work Regulations 1989. They must be able to interpret the legislation and assess the risks in respect of electrical equipment and appliances within their charge or which they are contracted to inspect and test. Training is therefore needed to understand sections 1 to 8 of this Code of Practice.

Managers must also maintain the records of inspections and tests of appliances and equipment and manage the re-inspection and re-testing at appropriate intervals as specified. It is also part of their duties to interpret the recorded results and to take appropriate actions regarding equipment or report to a more senior person within their organisation.

Competence to interpret recorded results is usually achieved by appropriate training and experience.

9.4 Inspectors

Those carrying out inspection and testing must be competent to undertake the inspection and, where appropriate, testing of electrical equipment and appliances having due regard to their own safety and that of others.

The tester must have an understanding of the modes of electrical, mechanical or thermal damage to electrical equipment and appliances and their flexes which may be encountered in any environment.

Training must include the identification of equipment and appliance types to determine the test procedures and frequency of inspection and testing. Persons testing must be familiar with test instruments used and in particular their limitations and restrictions so as to achieve repeatable results without damaging the equipment or appliance.

The importance must be stressed of recording inspection and test results, labelling and reporting to managers for action on defects, trends or changes in their assessment of risk.

10. TEST INSTRUMENTS

In order to carry out the prescribed tests which are detailed subsequently in the text, suitable test equipment, together with associated probes and leads, will be required.

All test instruments should be safe. The current safety standard is BS EN 61010 Safety requirements for electrical equipment for measurement, control, and laboratory use.

All new equipment should comply with this or an equivalent standard. This does not imply that existing equipment pre-dating this standard is unsafe.

BEFORE USING TEST INSTRUMENTS USERS MUST ALWAYS READ THE ACCOMPANYING OPERATING INSTRUCTIONS AND FOLLOW THE ADVICE GIVEN.

Test probes and leads, in particular those used to apply or measure voltage over 50 V a.c. and 100 V d.c. should comply with HSE Guidance Note GS 38.

Generally, portable appliance testers offer the most convenient means of providing the required test facilities but this does not preclude the use of suitable individual general test instruments. The following comments are not comprehensive, but are included to provide general guidance.

10.1 Portable appliance test instruments

Portable appliance test instruments (PATs) provide the following facilities:

(a) earth continuity with one or more pre-set test currents up to a maximum of around 25 A

(b) insulation resistance, generally by the applied voltage method using a test voltage of 500 V d.c.

These testers may offer additional test facilities, such as:-

(c) dielectric strength testing.

(d) insulation resistance measurement by the earth leakage method

(e) load test

(f) earth continuity using a low value of current, typically 100 mA, known as a 'soft test'

10.2 Low resistance ohmmeters (for earth continuity testing)

Earth continuity testing may in certain circumstances (see paragraph 15.4 of Part 2) be carried out by a low resistance ohmmeter. This may be either a specialised low resistance ohmmeter, or the continuity range of a combined insulation and continuity tester. The test current may be a.c. or d.c., but should be derived from a source with an open-circuit voltage of not less than 100 mV and no greater than 24 V. The test current should be within the range 20 to 200 mA.

The resolution at low scale values should be 0.01 ohm. A basic instrument accuracy within 2 per cent should be adequate. Field effects contributing to in-service errors are contact resistance, test lead resistance, a.c. interference and thermocouple effects in mixed metal systems.

Whilst contact resistance cannot be eliminated with two terminal testers, and can introduce errors up to 0.01 ohm, the effects of lead resistance can be eliminated by measuring this prior to a test, and subtracting the resistance from the final value. Test lead resistance may alternatively be "nulled" prior to the test, where the test instrument includes this facility.

Interference from an a.c. source cannot be eliminated. Thermocouple effects, if any, can be eliminated by reversing the test probes and averaging the resistance readings taken in each direction.

10.3 Insulation resistance ohmmeters (Applied voltage method)

The test voltage recommended in this Code of Practice for equipment is 500 V d.c.

The instrument used should be capable of maintaining the test voltage when applied to the insulation of the equipment under test.

10.4 Dielectric strength testing

Dielectric strength testing is not normally carried out during in-service testing.

This is normally carried out by the manufacturer on a complete appliance after assembly.

10.5 Earth leakage measurement (Protective conductor and touch current measurement)

This facility may be available on portable appliance test instruments.

10.6 Instrument accuracy

The accuracy of a test instrument should be verified and recorded annually or in accordance with the manufacturer's instructions.

PART 2 INSPECTION AND TESTING
(including user checks)

11. EQUIPMENT CONSTRUCTIONS

There are a number of basic equipment constructions that are referred to in all standards for electrical equipment and in this Code of Practice. They are important because they determine how the user is protected against electric shock and describe tests appropriate to apply when assessing safety. Appliances are often not what they appear, so a number of typical constructions are shown.

11.1 Class I

This is equipment (including appliances and tools) where protection against electric shock is achieved by:

(a) using basic insulation and also

(b) providing a means of connecting to the protective earthing conductor in the fixed installation wiring, those conductive parts (metal parts) that could otherwise assume hazardous voltages if the basic insulation should fail.

Class I equipment may have parts with double insulation or reinforced insulation or parts operating in extra-low voltage circuits.

Where equipment is intended to be used with a power supply flexible cable, there must be a protective earthing conductor incorporated in the cable.

Note: Class I equipment relies for its safety upon a satisfactory earth in the fixed installation and an adequate connection, usually via a flexible cable, to it.

(i) Class I typical construction showing basic insulation and earthed metal.

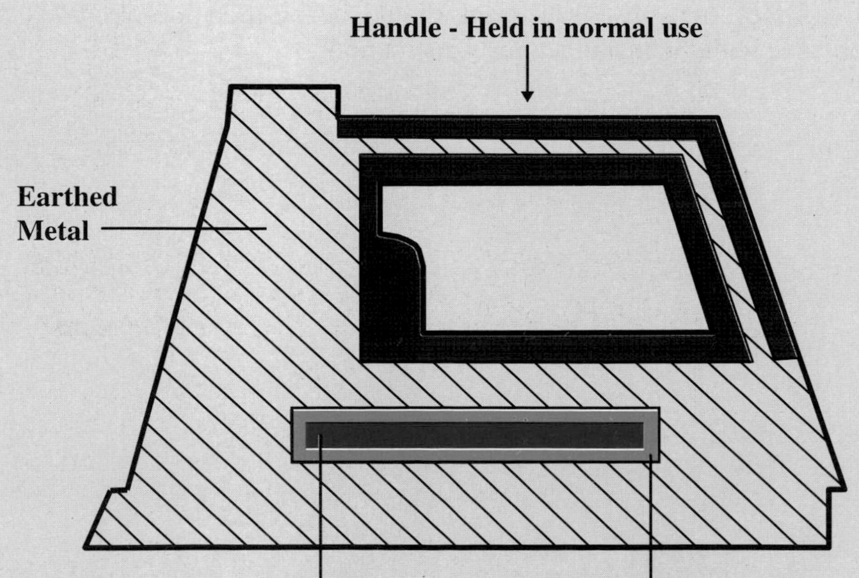

(ii) Class I construction showing the use of air as a basic insulation medium.

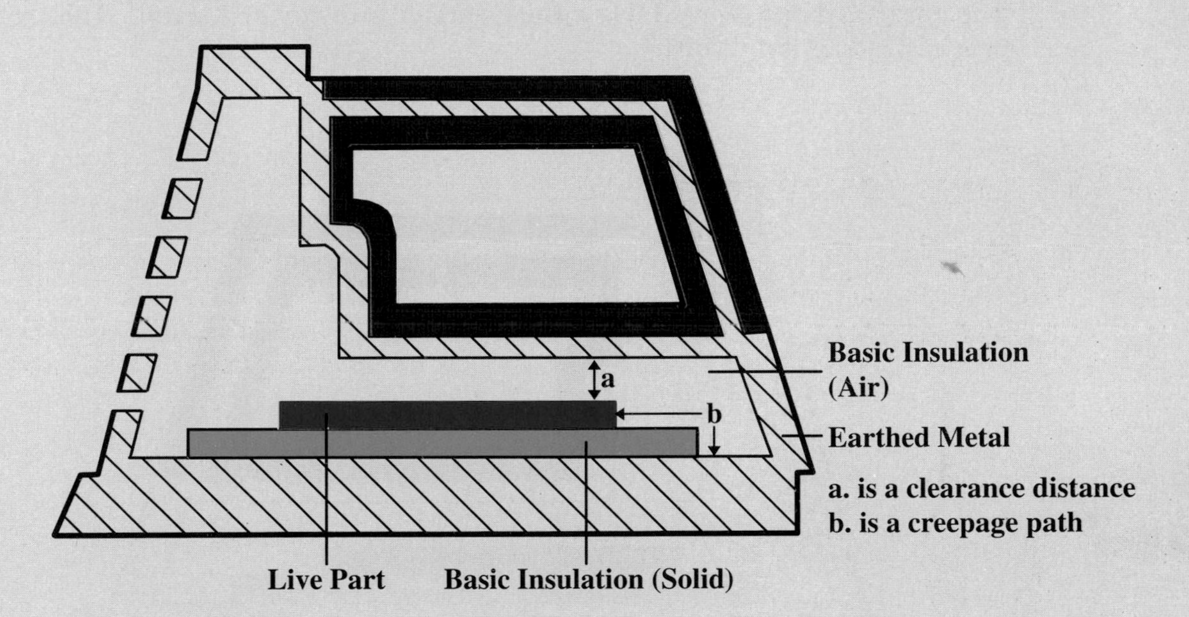

(iii) Class I construction incorporating unearthed metal separated from live parts by basic and supplementary insulation.

Unearthed metal may be encountered in Class I appliances as shown in drawings (iii) and (iv).

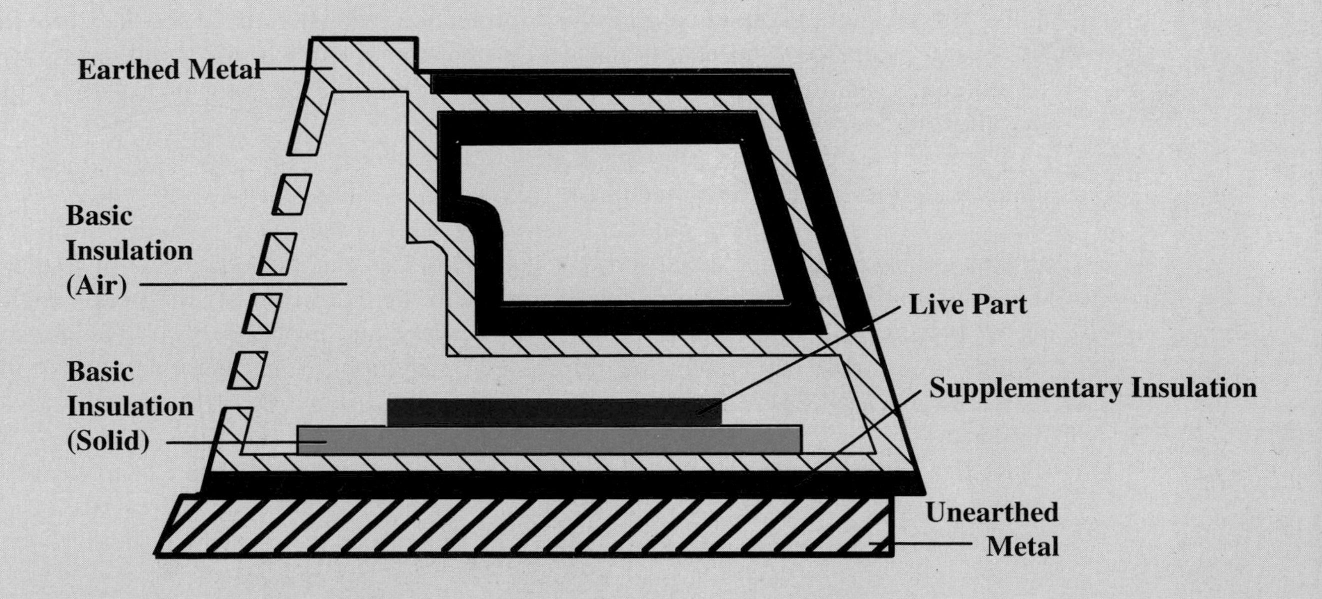

(iv) Class I construction incorporating unearthed metal separated from live parts by basic insulation and earthed metal.
The unearthed metal may be in casual contact with earthed metal. This can give misleading test results.

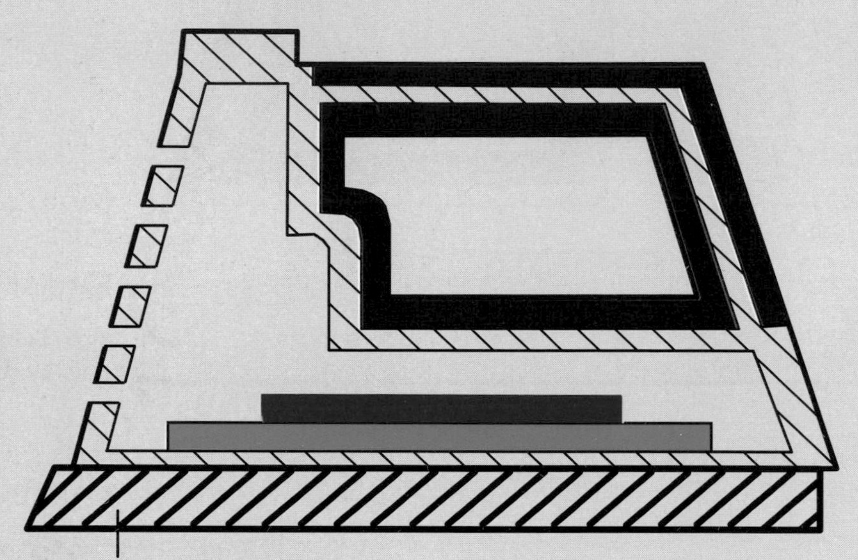

11.2 Class II

Class II equipment is equipment in which protection against electric shock is provided by basic insulation and additional safety precautions such as supplementary or reinforced insulation. There is no provision for protective earthing **or reliance upon installation conditions.**

Class II equipment may be of one of the following basic types:

Equipment having a durable and substantially continuous electrical enclosure of insulating material which envelops all conductive parts with the exception of small parts such as name plates, drill chucks, screws and rivets which are isolated from live parts by insulation at least equivalent to reinforced insulation. Such equipment is called insulation-encased Class II.

Class II equipment may have a substantially continuous metal enclosure with double or reinforced insulation used throughout. Such equipment is called metal-cased Class II equipment.

(i) **Class II equipment with a substantial enclosure of insulating material comprising basic and supplementary insulation.**

Live Part

Supplementary Insulation

Basic Insulation

(ii) **Class II equipment with a substantial enclosure of reinforced insulating material.**

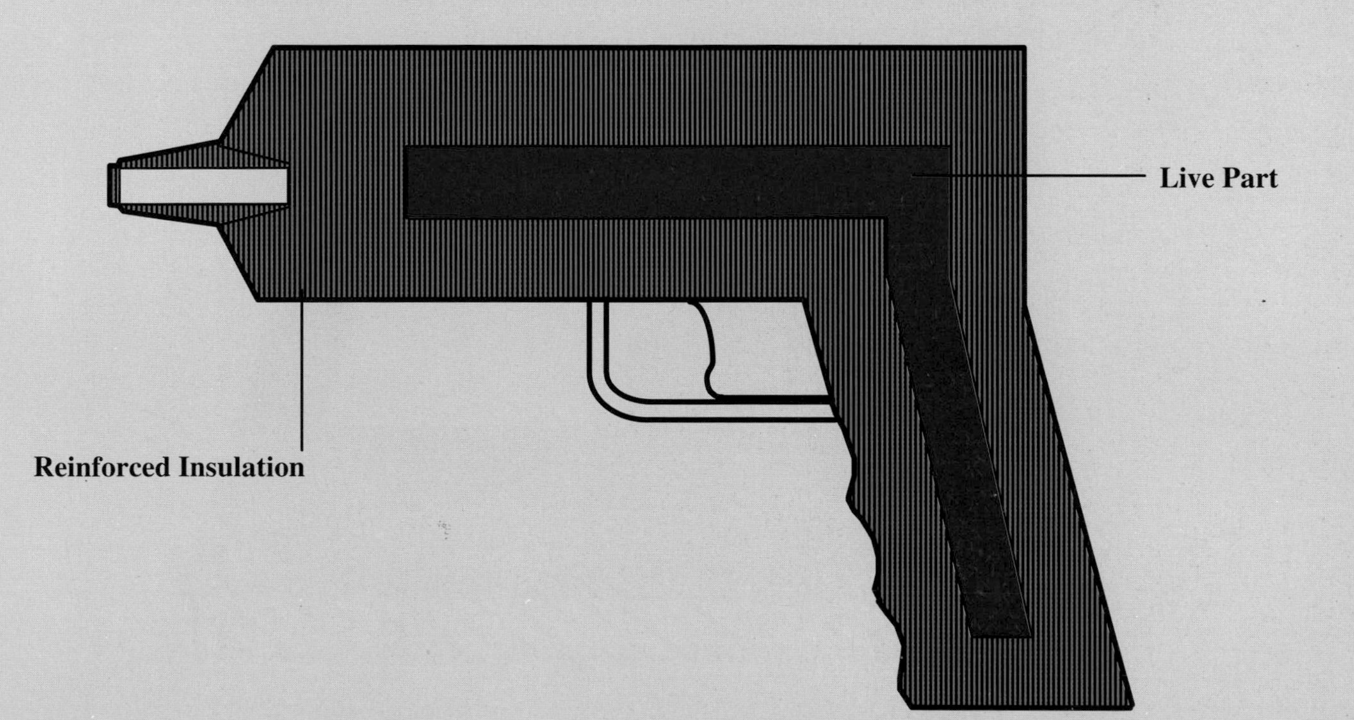

(iii) Class II equipment with a substantial enclosure of insulating material - the insulation construction includes air.

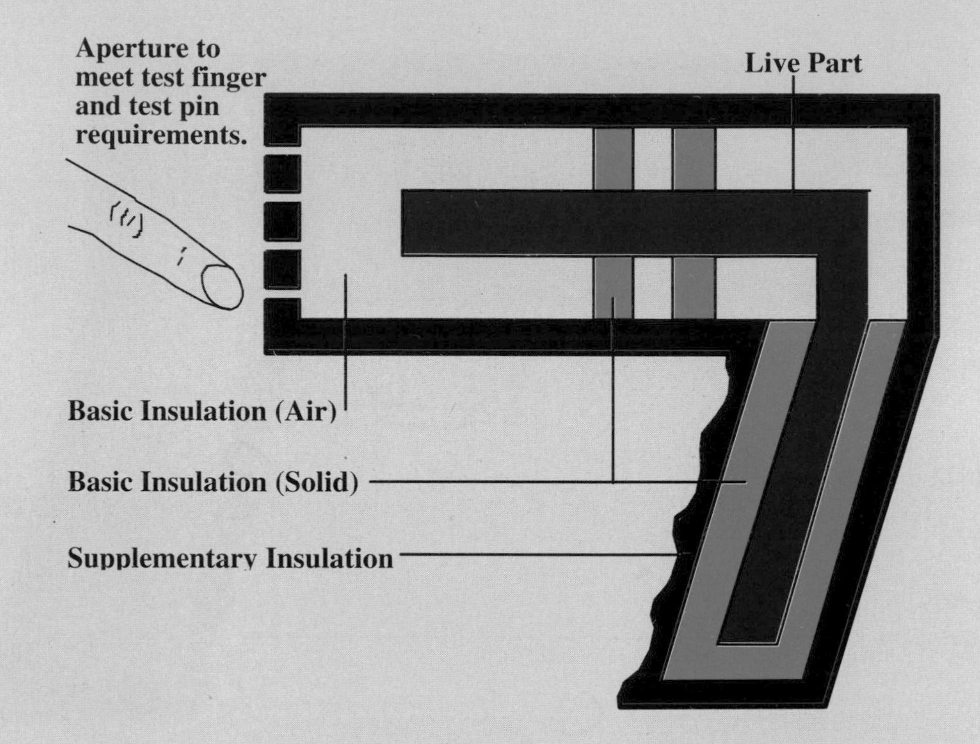

(iv) Class II equipment with unearthed metal in the enclosure, separated from live parts by basic and supplementary insulation.

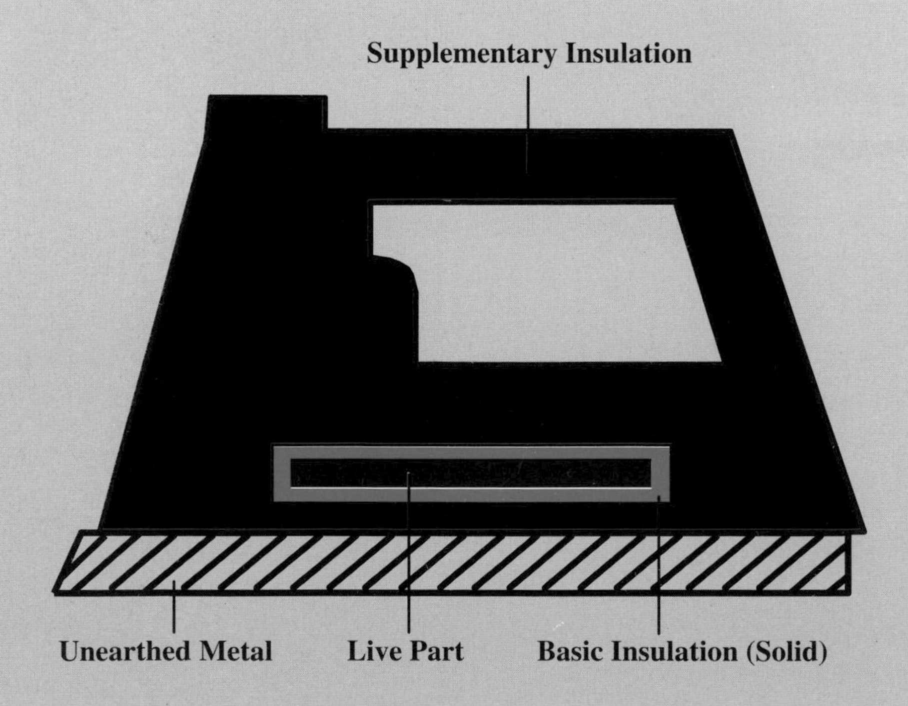

(v) Class II equipment with unearthed metal separated from live parts by reinforced insulation.

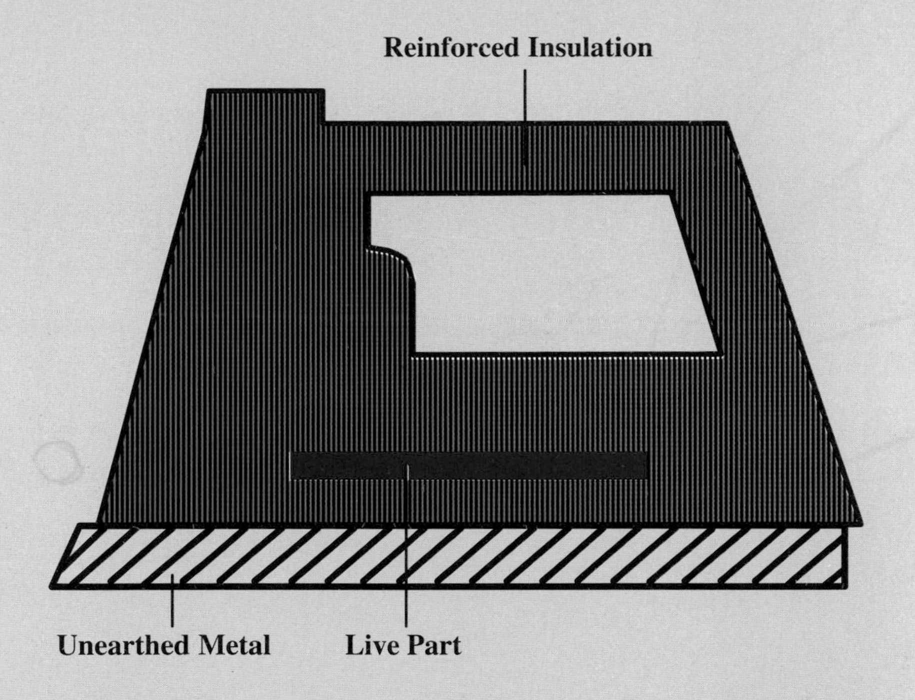

(vi) Class II equipment with unearthed metal separated from live parts by basic and supplementary insulation including air gaps

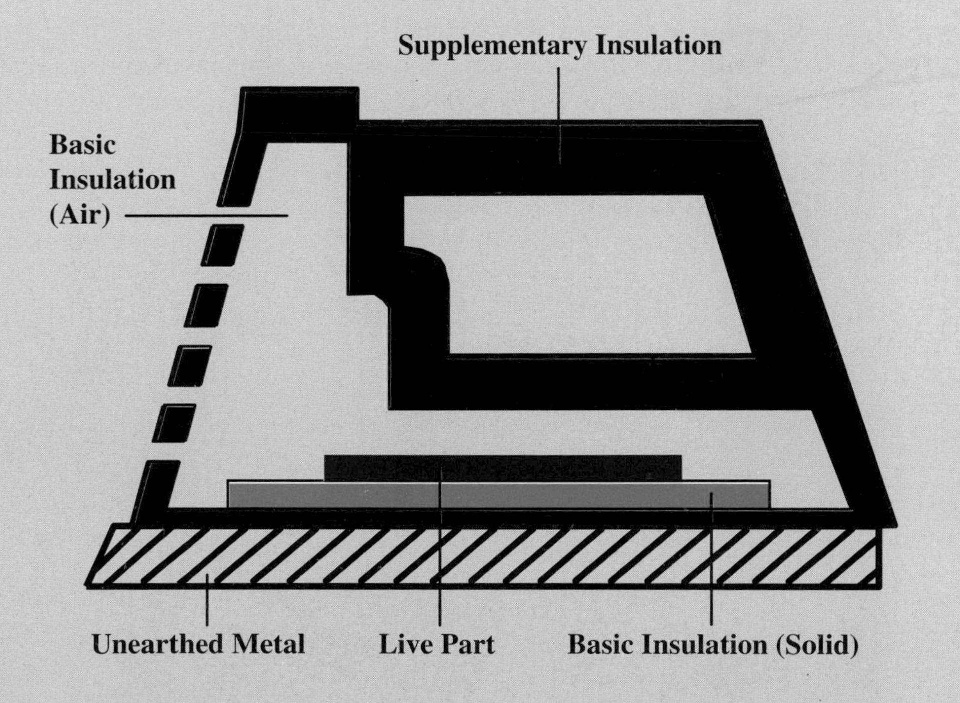

Class II equipment should be identified with the Class II construction mark:

11.3 Class III

Class III equipment relies for protection against electric shock on supply from a SELV source.

Note
SELV is described as Safety extra-low voltage in appliance standards e.g. BS EN 60335, and separated extra-low voltage in installation standards e.g. BS 7671

The Class III construction mark is as follows:

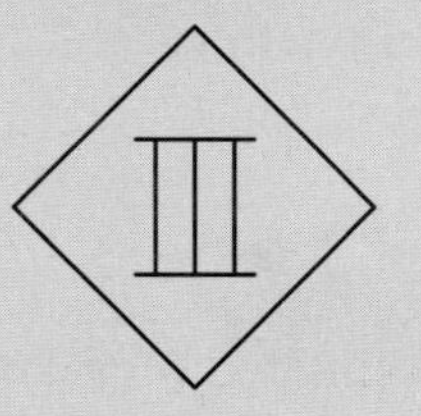

SELV sources will not exceed 50 V a.c. and in many installations will be required to be below 24 or 12 V. SELV systems require specialist design and there must be no earth facility in the distribution of a SELV circuit nor on the appliance or equipment.

Class III equipment must be supplied from a safety isolating transformer to BS EN 60742 or BS EN 61558-2-6.
The safety isolating transformer will have the following identification mark upon it:

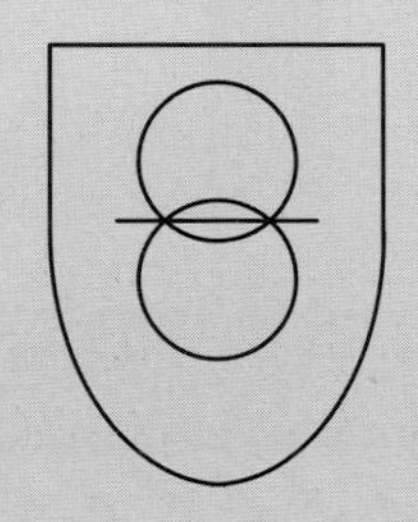

The output winding of the safety isolating transformer is electrically separated from the input winding by insulation at least equivalent to double insulation or reinforced insulation.

Class I and Class II appliances are equipment which is commonly encountered. Class III is used in the form of SELV lighting for shops and offices.

11.4 Class 0 and 0I

For completeness, Class 0 and Class 0I appliances are described below. Their use is allowed only in very specific locations, and they should not be used in the normal business or domestic environment.

Class 0 appliance

A Class 0 appliance is an appliance in which protection against electric shock relies upon basic insulation. This means that there is no provision for the connection of accessible conductive parts, if any, to the protective conductor in the fixed wiring of the installation. Reliance in the event of failure of the basic insulation is placed upon the environment in which the equipment is installed.

Class 0I appliance

A Class 0I appliance has at least basic insulation throughout, and is provided with an earthing terminal, but with a power supply cord without a protective earthing conductor and fitted with a plug without an earthing contact which cannot be introduced into a socket-outlet with an earthing contact.

12. TYPES OF INSPECTION AND TESTING

Three types of inspection and testing are described in Table 1 of Part 1, SUGGESTED INITIAL FREQUENCY OF INSPECTION AND TESTING OF EQUIPMENT, as follows:

(a) user checks (no record is taken if the equipment is found to be satisfactory)

(b) formal visual inspections (recorded)

(c) combined inspection and testing (recorded).

13. USER CHECKS

User checks are an important safety precaution. Many faults can be determined by a visual inspection. The user is the person most familiar with the equipment and may be in the best position to know if it is in a safe condition and working properly. No record is made of user checks unless some aspect of the equipment is reported to be unsatisfactory. Advice on the frequency of user inspections is given in Table 1 of Part 1.

The user checks should proceed as follows:

(a) consider whether he/she (the user) is aware of any fault in the equipment and whether it works properly

(b) disconnect the equipment if appropriate, as described in paragraph 14.4.

(c) Inspect the equipment, in particular looking at:

  (i) the flex - is it in good condition? Is it free from cuts, fraying and damage? Is it in a location where it could be damaged, is it too long, too short or in any other way unsatisfactory? Does it have any joints, which may render it unsuitable for use?

  (ii) the plug (where fitted) - is the flexible cable secure in its anchorage? Is it free from any sign of overheating? Is it free from cracks or damage?

  (iii) the socket-outlet or flex outlet - is there any sign of overheating? Is it free from cracks or other damage?

  (iv) the appliance - does it work? Does it switch on and off properly? Is it free from cracks, chemical or corrosion damage to the case, or damage which could result in access to live parts? Can it be used safely?

  (v) the environment - is the equipment suitable for its environment?

  (vi) suitability for the job - is the equipment suitable for the work it is required to carry out?

(d) Take action on faults/damage.

Faulty equipment must be:

  (i) switched off and unplugged from the supply
  (ii) labelled to identify that it must not be used
  (iii) reported to the responsible person.

Note: If equipment is found to be damaged or faulty, an assessment must be made by a responsible person as to the suitability of the equipment for the use/location. If the responsible person concludes that the equipment is unsafe, it must be immediately removed from use. Frequent inspections and tests will not prevent damage occurring if the equipment is unsuitable. Replacement by suitable equipment is required.

14. FORMAL VISUAL INSPECTIONS

Formal visual inspections should be carried out only by persons competent to do so.

The results of a formal visual inspection must be noted on a form such as Form Vb in Appendix V. Advice on the frequency of recorded inspections is given in Table 1 of Part 1. The following must be considered when carrying out a formal recorded inspection of equipment.

14.1 The environment

When the work environment is harsh or hazardous, e.g. if the equipment is exposed to mechanical damage, the weather, natural hazard, high or low temperatures, pressure, wet, dirty or corrosive conditions, flammable or explosive substances, particular care needs to be taken when selecting the equipment and assessing the frequency of inspection and testing. In hazardous and particularly difficult environments, specialist advice needs to be taken and reference must be made to British Standards and Health and Safety Executive Guidance, e.g. the guidance on Regulation 6 in the Memorandum of Guidance on the Electricity at Work Regulations 1989 - see Appendix II.

14.2 Good housekeeping

Check that the equipment is installed and operated in accordance with the manufacturers' instructions. Notwithstanding the manufacturers' instructions, the following are examples of items which should be checked:

(a) cables are not located where they are likely to be damaged, e.g. trodden upon or snagged, or create trip hazards

(b) means of disconnection/isolation from the mains supply are readily accessible

(c) space around the equipment is adequate for ventilation and cooling and equipment ventilation openings are not blocked

(d) cups, plants and work material are not placed where their contents could spill into equipment

(e) equipment is not positioned so close to walls and partitions that the cord is forced into a tight bend as it exits the equipment (this may also indicate inadequate spacing for ventilation and cooling)

(f) the equipment is operated with protective covers in place and doors closed

(g) there is no indiscriminate use of multiway adaptors and trailing socket-outlets

(h) there are no unprotected cables run under carpets.

14.3 Suitability of the equipment

If the inspector or tester considers that the equipment being inspected and/or tested is not suitable either for:

(a) the environment or

(b) the nature of the work being undertaken,

this should be recorded on the documentation and brought to the attention of the responsible person. It is appropriate to quote from Regulation 6 of the Electricity at Work Regulations 1989:-

> *"Electrical equipment which may reasonably foreseeably be exposed to:*
>
> *(a) mechanical damage;*
>
> *(b) the effects of weather, natural hazards, temperature or pressure;*
>
> *(c) the effects of wet, dirty, dusty or corrosive conditions; or*
>
> *(d) any flammable or explosive substance, including dust, vapours or gases*
>
> *shall be of such construction or as necessary protected as to prevent, so far as is reasonably practicable, danger arising from such exposure".*

Similar requirements are found in Regulation 4 of the Provision and Use of Work Equipment Regulations 1998, which also requires the employer to ensure that equipment is only used for operations and under conditions for which it is suitable.

14.4 Disconnection of equipment

The means of isolation from the electricity supply must be readily accessible to the user, i.e. in normal circumstances it must be possible to reach the plug and socket without difficulty. In general, the inspector will determine whether there is a means for switching off the electricity:

(a) for normal functional use
(b) in emergency
(c) to carry out maintenance.

Where possible the equipment must be isolated from the supply. This will be simple to achieve when the equipment is connected via a plug and socket. However, some equipment may be connected to the supply by other means such as an isolator or connection unit, where isolation from the supply can be achieved only by switching OFF or by removing the fuse. Great care should be taken when carrying out a visual inspection of equipment which does not have a visible means of isolation.

Before isolating business, telecommunication and other equipment from the supply, the permission of the responsible person will need to be obtained, otherwise disruption or serious loss to the business may result. Similarly, the permission of the responsible person will need to be obtained before disconnecting communication links.

It should be noted that business equipment may need to be powered down before isolation. Specific procedures may need to be followed to prevent damage or loss of data etc.

Equipment supplied via an uninterruptible power supply (or other standby supply) must be isolated from its standby source before the inspection commences.

14.5 The condition of the equipment

Before inspecting the equipment ask the users whether they are aware of any faults and whether it works properly, and proceed accordingly. The user is familiar with the equipment and may be aware of intermittent faults.

The following items need to be inspected:

(a) the flexible cable - is it in good condition? Is it free from cuts, fraying and damage? Is it in a location where it could be damaged or cause a trip hazard? Is it too long, too short or in any other way unsatisfactory?

(b) the socket-outlet (if known) or flex outlet - is there any sign of overheating? Is it free from cracks and other damage?

(c) the appliance - does it work? Does it switch on and off properly? Is it free from cracks or damage to the case or damage which could result in access to live parts? Can it be used safely?

(d) the standard plug

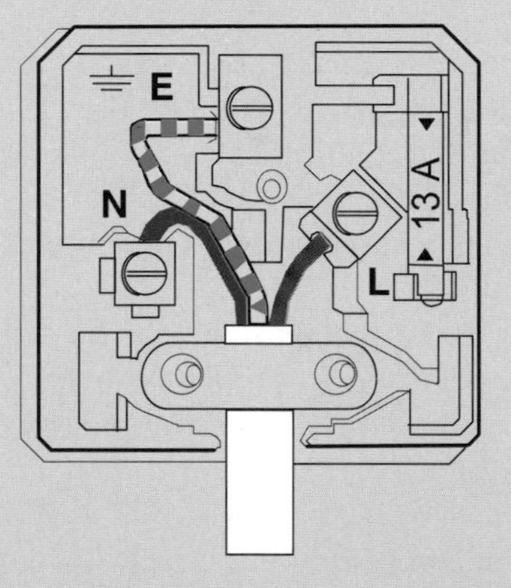

L brown (or red)*
N blue (or black)*
E green-and-yellow (or green)*

* May be found fitted to old appliances.

Some of the following checks may not be possible for equipment fitted with a non-rewirable plug

(i) check that detachable power supply cords to Class I equipment incorporate a continuous protective conductor

(ii) look for signs of overheating - this may be caused by a fault in the plug (e.g. a loose connection) **or by a faulty socket-outlet (or connection)**

(iii) remove the cover of the plug. Check that the flexible cable is properly secured in the cord anchorage - gripping the sheath so that there is no strain on the cable cores or the terminations

(iv) if the plug is of the non-rewirable type, the cable grip should be tested by firmly pulling and twisting the cable. No movement should be apparent

(v) check that the cable core terminations are tight, the plug is correctly connected, there is no excessive removal of insulation, that there are no loose strands and the cable cores are not strained

(vi) the fuse should be securely gripped, and should not show any signs of overheating. Check that the fuse is to BS 1362 and is approved - an ASTA mark shows that it has been approved for safety. Check the rating of the fuse - most appliances up to about 700 W should have a 3 A fuse fitted (red). For appliances over about 700 W fit a 13 A fuse (brown). Non-rewirable plugs will have the appropriate fuse rating marked on them

(vii) when replacing the plug cover check that it fits properly and will not come loose during use

(viii) check the flexible cable connections and anchorage at the equipment, if practicable

The following also need consideration:

(e) suitability of the equipment - it is an essential part of any equipment inspection to confirm that the equipment is suitable for the work, the environment and method of use - see paragraphs 14.1, 14.2 and 14.3.

(f) users - are they satisfied that the equipment works properly?

15. COMBINED INSPECTION AND TESTING

When electrical testing is required it should be performed by a person who is competent in the safe use of the test equipment and who knows how to interpret the results obtained. This person must be capable of inspecting the equipment and, where necessary, dismantling it to check the cable connections. Care must always be exercised when conducting tests. **Remember, inappropriate tests can damage equipment.**

If equipment is permanently connected to the fixed installation, e.g. by a flex outlet or other accessory, the accessory will need to be detached from its box or enclosure so that the connections can be inspected. Such work should only be carried out by a competent person.

Guidance on the initial frequency of inspection and testing is given in Table 1 of Part 1.

Before inspection and testing is carried out, the tester should obtain a copy of the previous test results so that any deterioration can be assessed and advice given accordingly.

15.1 Preliminary inspection

Before equipment is tested it is necessary to carry out a preliminary visual inspection. This visual inspection is the most important activity to be carried out on a piece of equipment. Formal testing with instruments often will not indicate a failure which is apparent on inspection, for example a damaged case exposing live parts will often not be apparent from an insulation test.

Preliminary inspection procedure:

(a) determine whether the equipment can be disconnected from the supply and disconnect if, and only if, permission is received (see paragraph 14.4). If permission is not received to disconnect the supply do not proceed with any tests. Record that the equipment has not been inspected and label accordingly

(b) disconnect business equipment from communication links after, and only after, receiving permission. Do not test equipment connected to communication links. Care should be taken with optical fibre systems to avoid exposure to invisible IR radiation. Do not hold up optical fibre links to the eye and ensure that disconnected fibre ends are protected with their resident dust-caps

(c) thoroughly inspect the appliance for signs of damage, as for a recorded inspection (paragraph 14.5)

(d) inspect the flexible cable for damage throughout its length by visual inspection and touch

(e) inspect the plug as for a recorded inspection (paragraph 14.5(d)), and check that the plug is suitable for the application. A resilient plug marked BS 1363 A may be necessary if the plug is subjected to harsh treatment e.g. vacuum cleaners, lawnmowers and extension leads

(f) assess if the equipment is suitable for the environment

(g) inspect the socket-outlet or flex outlet, as for a recorded inspection (paragraph 14.5(b)).

All standard 13 A plugs now sold are required by law (The Plugs and Sockets etc. (Safety) Regulations 1994) to conform to BS 1363 which requires pins to be sleeved. The legislation is not retrospective in that it does not apply to old plugs already in use, but such plugs should not be reused (i.e. not refitted to new cords).

15.2 Test procedures

Equipment that is supplied by a plug and socket connection can be readily tested by a dedicated portable appliance test instrument, by unplugging the equipment to be tested and plugging it into the dedicated test equipment.

Equipment that is permanently connected to a flex outlet type of accessory can more easily be tested using an insulation/continuity tester with the test leads connected directly to the accessory terminals. The supply to the accessory must be isolated before the testing commences.

15.3 In-service tests

In-service testing must be preceded by a preliminary inspection as described in paragraph 15.1. Testing will involve the following:

(a) earth continuity tests

(b) insulation resistance testing (this may be substituted by touch current measurement where insulation resistance testing is not appropriate)

(c) functional checks, to be carried out in the above sequence.

Some electrical test devices apply tests which are inappropriate and may even damage equipment containing electronic circuits, possibly causing degradation to safety. In particular, whilst this Code of Practice includes insulation resistance tests, equipment should not be subjected to dielectric strength testing (known as hi-pot testing or flash testing) because this may damage insulation and may also indirectly damage low voltage electronic circuits unless appropriate precautions are taken.

15.4 Earth continuity testing

This test can only be applied to Class I equipment or cords, that is equipment which:

(a) relies on a connection with earth for its safety (protective earthing)

and/or

(b) needs a connection with earth for it to work (functional earthing).

If protective earthing is being provided, as is likely for household appliances, tools, luminaires etc., the earth continuity test is most important, as the safety of the appliance depends upon a good connection with the earth of the fixed electrical installation.

Either of the following tests may be carried out:

(i) A continuity measurement with a short-circuit test current within the range 20 mA to 200 mA, carried out while flexing the flexible cable at the points of entry to the equipment and the plug, applied between accessible earthed metal parts of the equipment and the earth pin of the plug (or the earthing terminal of the fixed wiring supply). The terminations are inspected for signs of deterioration, poor contact, corrosion etc.

or

(ii) A continuity measurement with a test current not less than 1.5 times the rating of the fuse and no greater than 25 A for a period of between 5 and 20 seconds, and including an inspection of the flexible cable terminations at the plug/flex outlet and at the equipment.

The resistance shall not exceed the following:-

(0.1 + R) ohm where R is the resistance of the protective conductor of the supply cord, and

0.1 ohm for appliances without a supply cord.

Notes:

(1) Some equipment may have accessible metal parts which are earthed only for functional/screening purposes with protection against electric shock being provided by double or reinforced insulation. **It is very important that these non-safety earthed metal parts are not subjected to the above test (ii) otherwise damage may result.** Connections may be checked using a low current continuity tester as in test (i).

(2) Care should be taken to ensure that the contact resistance between the tip of the test probe and the metal part under test does not influence the test result.

(3) The test should only be carried out for the duration necessary for a stable measurement to be made, and to allow time for flexing of the cable.

(4) If the resistance of the protective conductor of the supply cord cannot easily be measured, Table VI (Appendix VI) provides nominal cable resistances per metre length for various types of cable. The supply cord cross-sectional area must first be identified and the length measured. The resistance of the protective conductor can then be calculated.

(5) Some portable appliance testers with go/no-go indication may fail cord connected appliances with earth continuity resistance exceeding 0.1 ohm. If it is not possible to re-programme the appliance tester it will be necessary for a measurement of the actual resistance to be made with another instrument.

Supply cords must not be extended by taped joints and should not exceed the length allowed by the equipment standard. Where extension leads are necessarily used, guidance given in paragraph 15.10 should be followed.

15.5 Insulation resistance testing

For some types of appliance, insulation resistance is generally checked by applying a test voltage and measuring resistance. This test may not always be suitable, and the touch current measurement of paragraph 15.6 may be a more appropriate alternative.

Appliances should not be touched whilst carrying out insulation resistance tests as exposed metalwork may reach the test voltage which, whilst not dangerous, could be uncomfortable and risk causing injury by involuntary movement.

It is important that the live conductors (phase and neutral) are connected together for this test. This is best achieved either by using special test equipment or by using a special test socket with the phase and neutral connected together. It is important that equipment is not returned to service with any phase-neutral connections still in place. It is therefore recommended that functional tests are carried out last.

Before the test, the suitability of fuses should be checked and power switches put in the ON position. During the test all covers should be in place. The test is carried out between live conductors, i.e. phase and neutral, connected together and the body of the appliance.

The applied test voltage should be approximately 500 V d.c. The test instrument should be capable of maintaining this test voltage with a load resistance of 0.5 Megohm. Insulation resistance readings obtained should be not less than the values shown in Table 2 below:

TABLE 2 - INSULATION RESISTANCE READINGS

Appliance Class	Insulation Resistance
for Class I heating equipment with a rating $\geq$ 3 kW	0.3 megohm
All other Class I equipment	1.0 megohm
Class II equipment	2.0 megohm
Class III equipment	250 kilohm

Notes to Table 2:

(1) Heating and cooking appliances: the earth leakage of appliances with relatively highly loaded heating elements could be such that they may be unable to meet the insulation resistance requirements. This may particularly be the case when metal sheathed mineral-insulated elements are used. It may be necessary in some cases to switch on the appliance for a period of time to drive off absorbed moisture before commencing testing. Alternatively, the touch current measurement of paragraph 15.6 may be carried out.

(2) It is important to ensure that there is a good connection with earth, particularly for appliances with relatively high protective conductor current, otherwise electric shocks may be received from the appliance frame.

(3) This test should not be applied routinely to information technology equipment, unless it complies with BS EN 60950. Equipment not constructed to this standard may be damaged by this test.

(4) The phase and neutral of the equipment must be securely connected together while making this test. This is best achieved by using pre-wired automatically-configured test equipment or by plugging into a special test socket.

(5) For 3-phase equipment, all three phases and neutral (if applicable) must be linked together whilst making this test.

(6) Some equipment may have filter networks or transient suppression devices which could cause the insulation resistance to be less than specified. The manufacturer/ supplier must be consulted in these cases.

15.6 Protective conductor/ touch current measurement

Protective conductor/ touch current measurement is an alternative to the in-service insulation test for use if the insulation resistance test either cannot be carried out or gives suspect test results. The current is measured from live parts to earth for Class I equipment, or from live parts to accessible surfaces of Class II equipment.

The current is to be measured within 5 seconds after the application of the test voltage and must not exceed the values in Table 3.

For practical purposes the test voltage is the supply voltage.

TABLE 3 - MEASURED PROTECTIVE CONDUCTOR/ TOUCH CURRENT

Appliance Class	**Maximum Current note (1)**
Portable or hand-held Class I equipment	0.75 mA
Class I heating appliances	0.75 mA or 0.75 mA per kW, whichever is the greater, with a maximum of 5 mA
Other Class I equipment	3.5 mA
Class II equipment	0.25 mA
Class III equipment	0.5 mA

Notes to Table 3:

(1) The values specified above are doubled:

– if the appliance has no control device other than a thermal cut-out, a thermostat without an "off" position or an energy regulator without an "off" position.

– if all control devices have an "off" position with a contact opening of at least 3 mm and disconnection in each pole.

(2) Equipment with a protective conductor current designed to exceed 3.5 mA shall comply with the requirements of paragraph 15.12.

(3) The nominal test voltage is:

- 1.06 times rated voltage, or 1.06 times the upper limit of the rated voltage range, for appliances for d.c. only, for single-phase appliances and for three-phase appliances which are also suitable for single-phase supply, if the rated voltage or the upper limit of the rated voltage range does not exceed 250 V;
- 1.06 times rated line voltage divided by 1.73, or 1.06 times the upper limit of the rated voltage range divided by 1.73 for other three-phase appliances.

15.7 Functional checks

In its simplest form a functional check is simply a check to ensure that the appliance is working properly.

The use of more sophisticated instruments may permit load testing, which is an effective way of determining whether there are certain faults in appliances. It is particularly useful for heating appliances and will identify whether one or more elements are open-circuit.

15.8 Damaged/faulty equipment

If equipment is found to be damaged or faulty on inspection or test, an assessment must be made by a responsible person as to the suitability of the equipment for the use/ location. Frequent inspections and tests will not prevent damage to equipment. If it is unsuitable it must be replaced by suitable equipment.

Any items found to be faulty or defective by the tester should be brought to the attention of the responsible person. Similarly, items of equipment on the register which the inspector is unable to locate should also be brought to the attention of the responsible person.

15.9 Appliance cord sets

Appliances with detachable power supply flexes (appliance-couplers) should be tested with the cord set plugged into the appliance. It is recommended that cord sets be labelled and tested separately from the appliance as follows:

3-core cord sets as a Class I appliance
2-core cord sets as a Class II appliance

The following are applicable:

Visual inspection
Class I - earth continuity, polarity and insulation checks
Class II - polarity and insulation checks

15.10 Extension leads

Where extension leads are fitted with a standard 3-pin socket-outlet these should be tested as Class I appliances with the addition of a polarity check. Any such extension leads that are found to be without an earth wire should be marked as defective and removed from service.

The length of an extension lead should be checked to ensure that it is not so great that the appliance performance may be affected by voltage drop. Additionally, the length should not exceed the following:

core area	maximum length
1.25 mm^2	12 metres
1.5 mm^2	15 metres
2.5 mm^2	25 metres

2.5 mm^2 extension leads are too large for standard 13 A plugs, although they may be used with BS EN 60309 industrial plugs. Extension leads exceeding the above lengths should be fitted with a 30 mA RCD manufactured to BS 7071.

Cable reels must be used within their reeled or unreeled ratings as appropriate.

15.11 Microwave ovens

When inspecting and testing microwave ovens showing any signs of damage, distortion or corrosion should be rejected. Return damaged ovens to specialist repairers only.

A functional check to ascertain that opening of the door results in a reliable interruption of the oven power should be carried out.

15.12 High protective conductor currents

There are particular requirements in Section 607 of BS 7671, Requirements for Electrical Installation (IEE Wiring Regulations) for the earthing arrangements for equipment having high protective conductor_currents.

It should be noted that equipment with a protective conductor_current designed to exceed 3.5 mA shall:

(a) be permanently wired to the fixed installation, or
be supplied by an industrial plug and socket to BS 4343 (BS EN 60309-2) and

(b) have internal protective conductors of not less than 1.0 mm^2 cross-sectional area (see clause 5.2.5 of BS EN 60950), and

(c) have a label bearing the following warning or similar wording fixed adjacent to the equipment primary power connection (see clause 5.1.7 of BS EN 60950)

WARNING HIGH LEAKAGE CURRENT

Earth connection essential before connecting the supply

or

WARNING HIGH TOUCH CURRENT
Earth connection essential before connecting the supply

Further precautions need to be taken for equipment with a protective conductor current exceeding 10 mA, see Section 607 of BS 7671. Similarly, for final circuits supplying a number of socket-outlets, where it is known or reasonably to be expected that the total protective conductor current in normal service will exceed 10 mA, additional precautions need to be taken.

When appliances have high protective conductor currents, substantial electric shocks can be received from exposed-conductive-parts and/or the earth terminal if the appliance is not earthed. It is most important that appliances with such high protective conductor currents are properly connected with earth before any supply is connected.

15.13 Replacement of appliance flexes

The maximum lengths recommended for extension leads are not applicable to appliance flexes or cord sets.

For flexes to be protected by the fuse in a BS 1363 plug there is no limit to their length, providing their cross-sectional areas (csa) are as below:

Fuse rating	minimum flex csa
3 A	0.5 mm^2
13 A	1.25 mm^2

Other considerations such as voltage drop may limit flex lengths. Smaller csas than those given above are acceptable if flex lengths are restricted. However, for replacement purposes the above simplified guidance is appropriate.

15.14 Plug fuses

For the convenience of users, appliance equipment manufacturers have standardised on two plug fuse ratings (3 A and 13 A) and adopted appropriate flex sizes. For appliances up to 700 W a 3 A fuse is used, for those over 700 W a 13 A fuse is used.

The fuse in the plug is not fitted to protect the appliance, although in practice it often does this. Appliances are generally designed to European standards for use throughout Europe. In most countries the plug is unfused. If the appliance needs a fuse to comply with its standard it must be fitted in the appliance. The fuse in the plug protects the flex against faults and can allow the use of a reduced cross-sectional area flexible cable. This is advantageous for such appliances as electric blankets, soldering irons and Christmas tree lights, where the flexibility of a small flexible cable is desirable.

PART 3 APPENDICES

APPENDIX I BRITISH STANDARDS

The following British Standards are relevant to equipment testing:

BS 1362 : 1973 (1992)	Specification for general purpose fuse links for domestic and similar purposes (primarily for use in plugs).
BS 1363-1 : 1995	13 A fused plugs.
BS 2754 : 1976 (1999)	Memorandum. Construction of electrical equipment for protection against electric shock.
BS 6500 : 2000	Flexible cords rated up to 300/500 V, for use with appliances and equipment intended for domestic, office and similar environments.
BS EN 50144	Safety of hand-held electric motor operated tools.
BS EN 60051-1	Direct-acting electrical measuring instruments and their accessories.
BS EN 60065	Safety requirements for mains-operated electronic and related apparatus for household and similar general use.
BS EN 60309	Plugs, socket-outlets and couplers for industrial purposes
BS EN 60335	Safety of household and similar electrical appliances.
BS EN 60598	Safety of luminaires
BS EN 60742	Isolating transformers and safety isolating transformers. Being replaced by BS EN 61558.
BS EN 60950	Specification for safety of information technology equipment including electrical business equipment. Identical to BS 7002.
BS EN 61010	Safety requirements for electrical equipment for measurement, control and laboratory use.
BS EN 61557	Electrical safety in low voltage distribution systems. Equipment for testing, measuring and monitoring of protective measures.
BS EN 61558	Safety of power transformers, power supply units and similar.

British Standards are available from:

		telephone numbers
BSI Standards		
389 Chiswick High Road	for general enquiries:	020 8996 9000
LONDON	for ordering & information:	020 8996 9001
W4 4AL.	for membership:	020 8996 7002

Fax: 020 8996 7001
E-mail orders@bsi.org.uk or info@bsi.org.uk
Web www.bsi.org.uk

APPENDIX II LEGAL REFERENCES AND NOTES

1. Health and Safety at Work etc. Act 1974 (see Note 1).

2. The Electrical Equipment (Safety) Regulations 1994 - Statutory Instrument No 3260.

3. The Plugs and Sockets etc. (Safety) Regulations 1994 - Statutory Instrument No 1768.

4. The Electricity at Work Regulations 1989 - Statutory Instrument No 635 (see Note 2).

5. The Management of Health and Safety at Work Regulations 1999 - Statutory Instrument No 3242 (see Note 3).

6. Provision and Use of Work Equipment Regulations 1998 - Statutory Instrument No 2306, as amended (see Note 4).

7. The Health and Safety (Display Screen Equipment) Regulations 1992 - Statutory Instrument No 2792.

8. The Lifting Operations and Lifting Equipment Regulations 1998 - Statutory Instrument No 2307.

9. Health and Safety Commission and Health and Safety Executive Guidance and Codes of Practice on meeting the requirements of the above:

   (i) L1 - A Guide to the Health and Safety at Work Act 1974.

   (ii) L22 - Approved Code of Practice & Guidance on the Provision and Use of Work Equipment Regulations 1998.

   (iii) L26 - Guidance on The Health and Safety (Display Screen Equipment) Regulation 1992.

   (iv) HSR25 - Memorandum of Guidance on the Electricity at Work Regulations 1989.

   (v) L21 - Approved Code of Practice & Guidance on the Management of Health and Safety at Work Regulations 1999.

Items 1 to 8 are available from The Stationery Office.

The Stationery Office	Telephone:	0870 600 5522
The Publications Centre	Fax:	0870 600 5533
51 Nine Elms Lane	E-mail	bookenquiries@theso.co.uk
LONDON	Web	www.ukstores.com
SW8 5DT.		

Items 9(i) to 9(v) are available from HSE Books.

PO Box 1999	Telephone:	01787 881165
SUDBURY	Fax:	01787 313995
Suffolk	E-mail	mailhsebooks@prolog.uk.com
CO10 6FS.	General web	www.hse.gov.uk
	Order via web	www.hsebooks.co.uk

HSE publications are also available through good booksellers.

Note 1: Health and Safety at Work etc. Act 1974

Section 2 of the Health and Safety at Work etc. Act 1974 puts on employers a general duty of care to their employees. Specifically:

"it shall be the duty of every employer to ensure, so far as is reasonably practicable, the health, safety and welfare at work of all his employees.

(2) Without prejudice to the generality of an employer's duty under the preceding sub-section, the matters to which that duty extends include in particular -

(a) the provision and maintenance of plant and systems at work that are, so far as is reasonably practicable, safe and without risks to health; ....".

Employers also have general duties to persons other than their employees as described in Section 3(1).

Section 7 of the Act imposes general duties on employees at work as follows:

It shall be the duty of every employee while at work:

(a) to take reasonable care for the health and safety of himself and other persons who may be affected by his acts or omissions at work; and

(b) as regards any duty or requirement imposed on his employer or any other person by or under any of the relevant statutory provisions, to co-operate with him so far as is necessary to enable that duty or requirement to be performed or complied with.

The Health and Safety at Work etc. Act 1974 is all-embracing, requiring all those concerned with an undertaking to do all that is reasonable to ensure the health and safety not only of persons directly employed, but other persons who may be associated with the work undertaken by the business. The requirements of the Act are general and widely applicable.

Note 2: The Electricity at Work Regulations 1989
Statutory Instrument No 635

The purpose of the Electricity at Work Regulations 1989 is to prevent death or injury to anyone from any electrical cause as a result of, or in connection with, work activities.

The Regulations impose duties upon employers, self-employed persons, employees while at work etc. Regulation 4 is so significant that it is worth quoting in full.

Regulation 4
Systems, work activities and protective equipment.

(1) *All systems shall at all times be of such construction as to prevent, so far as is reasonably practicable, danger.*

(2) *As may be necessary to prevent danger, all systems shall be maintained so as to prevent, so far as is reasonably practicable, such danger.*

(3) *Every work activity, including operation, use and maintenance of a system and work near a system, shall be carried out in such a manner as not to give rise, so far as is reasonably practicable, to danger.*

(4) *Any equipment provided under these Regulations for the purpose of protecting persons at work on or near electrical equipment shall be suitable for the use for which it is provided, be maintained in a condition suitable for that use, and be properly used.*

The Electricity at Work Regulations apply to all electrical equipment from battery hand-lamps to 400 kV transmission lines. The source of energy, the distribution systems and the current consuming equipment are all covered.

The Regulations include specific requirements for the strength, capability, suitability, insulation, earthing, protection against excess current and isolation of electrical systems. They also extend to work and equipment associated with such systems, such as near overhead power lines. There are requirements for precautions to be taken before working on equipment made dead, and for work on or near live or charged conductors.

Note 3: Management of Health and Safety at Work Regulations 1999 Statutory Instrument No 3242

Regulation 3 of the Management of Health and Safety at Work Regulations 1999 requires all employers and self-employed persons to assess the risks to workers and others who may be affected by their undertaking. Employers with five or more employees must also record the significant findings of that assessment. Guidance on the Management of Health and Safety at Work Regulations 1999 can be found in the Approved Code of Practice & Guidance published by the Health and Safety Commission. The Code has special legal status, if you follow its advice you will be doing sufficient to comply with the law in respect of matters on which the Code gives advice. Failure to do so can lead to successful prosecution for breach of Health and Safety law. The Code is essential reading, not just for employers and managers but for everyone with responsibility for other people in the work place - see Regulation 14(2).

This legislation requires an organisation to:

(1) assess the risks to all persons associated with their electrical equipment, identifying the significant risks, e.g. portable equipment used out of doors, and make a record of the assessment

(2) as appropriate, appoint a competent person to take responsibility for electrical maintenance including inspection and testing; ensuring that the person given this responsibility is competent in that he or she has sufficient training and experience, knowledge and other qualities to enable him or her properly to support the organisation.

Note 4: Provision and Use of Work Equipment Regulations 1998 Statutory Instrument No 2306 as amended.

General

The Provision and Use of Work Equipment Regulations 1998 (PUWER) cover most risks that can result from using work equipment. With respect to risks from electricity, compliance with the Electricity at Work Regulations 1989 (EAW Regulations) is likely to achieve compliance with PUWER Regulations 5-9, 19 and 22.

PUWER applies only to work equipment used by workers at work. This includes all work equipment (fixed, portable or transportable) connected to a source of electrical energy. PUWER does not apply to the fixed installations in buildings. The electrical safety of these installations is dealt with only by the EAW Regulations.

More detailed commentary on the individual requirements is given below.

Maintenance

Regulation 5(1) of PUWER requires employers to ensure that work equipment is maintained in an efficient state, in efficient working order and in good repair. This is consistent with the requirement in Regulation 4(2) of the EAW Regulations that systems shall be maintained to prevent danger.

Regulation 5(2) of PUWER requires employers to ensure that, where any machinery has a maintenance log, the log is kept up to date. Neither PUWER nor the EAW Regulations specifically require a maintenance log to be kept for machinery, but it is good practice to keep a log or record of maintenance. Records of visual checks of plugs, cables etc carried out by users of electrical machinery would not normally be considered to constitute a maintenance log. However, if formal records of maintenance of electrical machinery are kept in a maintenance log, then that log shall be kept up to date.

Inspection of work equipment that poses electrical risks

Regulation 6 of PUWER introduces a specific requirement for the inspection of work equipment. The Approved Code of Practice & Guidance L22, "Safe Use of Work Equipment Provision and Use of Work Equipment Regulations 1998" notes at paragraph 136 that:-

"Where the risk assessment under Regulation 3 of the Management of Health and Safety at Work Regulations 1999 has identified a significant risk to the operator or other workers from the installation or use of work equipment, a suitable inspection should be carried out." (Note that whilst the 1992 Regulations have since been replaced by the 1999 Regulations, the requirements for risk assessment have not changed.)

Inspection is required for work equipment:

a) after installation and before being put into service for the first time, or after assembly at a new site or in a new location;

b) at suitable intervals where it is exposed to conditions causing deterioration which is liable to result in dangerous situations and each time that exceptional circumstances which are liable to jeopardise the safety of the work equipment have occurred.

The risk assessment carried out under the Management of Health and Safety at Work Regulations 1999 will determine whether there are significant electrical risks which might justify an inspection under PUWER. If there are significant electrical risks then a competent person, probably the person who is trained to visually inspect and maintain the electrical equipment under the EAW Regulations, is required to inspect the equipment and record the results of the inspection under PUWER.

The above requirement of Regulation 6 in PUWER is consistent with the requirement in Regulation 4(2) of the EAW Regulations that all electrical systems shall be maintained to prevent danger, so far as is reasonably practicable. Guidance on the EAW Regulations has always stressed the importance of inspection, testing and record keeping if justified by the risk. This is consistent with the more specific requirements of PUWER.

Specific Risks

Regulation 7 of PUWER imposes requirements for the use, repair, modification, maintenance or servicing of work equipment in high risk situations, such as where a Permit to Work system may be appropriate. This is consistent with the requirements in Regulations 4(3), 13 and 14 of the EAW Regulations and guidance given previously on those Regulations.

Information and instructions

Regulation 8 of PUWER requires employers to provide to people who use work equipment adequate health and safety information and, where appropriate, written instructions about the equipment that they use. This duty extends to providing information and instructions about any electrical risks that are present. Details of what is needed are set out in the Regulation and the supporting guidance in the Approved Code of Practice L22. This is consistent with the requirements of Regulations 4(3) and 16 of the EAW Regulations.

Training

Regulation 9 of PUWER requires employers to ensure that all persons who use work equipment have received adequate training for purposes of health and safety. This is consistent with the requirements of Regulations 4(3) and 16 of the EAW Regulations.

Isolation from sources of energy

Regulation 19 of PUWER requires employers to ensure that, where appropriate, work equipment is provided with suitable means to isolate it from all sources of energy and to take appropriate measures to ensure that reconnection of any energy sources does not expose any person using work equipment to any risk to his(her) health and safety. This is consistent with the requirements of Regulations 12 and 13 of the EAW Regulations.

Maintenance operations

Regulation 22 of PUWER requires employers to take appropriate measures to ensure that work equipment is so constructed or adapted that, so far as is reasonably practicable, maintenance operations which involve a risk to health and safety can be carried out while the equipment is shut down. Where this is not reasonably practicable employers must ensure that maintenance operations can be carried out without exposing the person carrying them out to a risk to his(her) health and safety or appropriate measures must be taken for the protection of any person carrying out maintenance operations which involve a risk to his(her) health and safety. This is consistent with the requirements of Regulations 4(3), 13 and 14 of the EAW Regulations.

APPENDIX III HSE AND HSC GUIDANCE

Health and Safety Executive (HSE) and Health and Safety Commission (HSC) publications on Electrical Safety.

Guidance document	*Title*	*Electricity at Work Regulations particularly relevant*
PM 29	*Electrical hazards from steam/water pressure cleaners* ISBN 0 11 883538 6	4,6,7,8 and 10
PM 38	*Selection and use of electric handlamps* ISBN 0 11 883582 3	4,6,7,8,10 and 12
HSG 118	*Electrical safety in arc welding* ISBN 0 7176 0704 6	4,6,7,8,10 12,14 and 16
GS 6	*Avoidance of danger from overhead electric lines* ISBN 0 11 883045 7	4,14,15 and 16
HSG 141	*Electrical safety on construction sites* ISBN 0 11 883570 X	4-16 inclusive
HS(G)47	*Avoiding danger from underground services* ISBN 0 11 885492 5	4,14 and 16
GS 38	*Electrical test equipment for use by electricians* ISBN 0 11 883533 5	4,5,6,7,10,14 and 16
HS(G)85	*Electricity at Work - Safe Working Practices* ISBN 07176 0442 X	4,7,12,13,14 15 and 16
HS(G)38	*Lighting at work* ISBN 0 11 883964 0	4,13,14 and 15
	Safe use of electric induction furnaces Health and Safety Commission, Foundries Industry Advisory Committee Publication ISBN 0 11 883909 8	4,5,6,7,8,14,15 and 16
HS(G)107	*Maintaining portable and transportable electrical equipment* ISBN 0 7176 0715 1	4,6,7,8,9,10,11 and 12

The above are available from HSE Books:-

PO Box 1999	Telephone:	01787 881165
SUDBURY	Fax:	01787 313995
Suffolk	E-mail	maillhsebooks@prolog.uk.com
CO10 6FS.	General web	www.hse.gov.uk
	Order via web	www.hsebooks.co.uk

HSE publications are also available through good booksellers.

APPENDIX IV PRODUCTION TESTING

Manufacturers of equipment will carry out tests on equipment they manufacture to check that the equipment has been manufactured as intended.

Where equipment is approved by an approval body, e.g. BEAB, BABT, BSI, the approval body will agree with the manufacturer the inspection, test and quality assurance procedures to be taken to ensure that the products are safe, and within accepted manufacturing tolerance of the samples type-tested.

The following tests are typical of the electrical safety tests which may be required.

These tests are included for reference purposes only to assist in determining, in cases of doubt, whether the results of in-service tests are acceptable. In-service test results would not be expected to be 'better' than manufacturing test results. It is to be noted that in-service tests are not necessarily the same as manufacturing tests with respect to applied test voltages and currents.

Production tests may be relatively arduous and generally should only be applied to equipment in as-new condition. They are reproduced here as they may well be useful for those testing as-new equipment or repaired equipment if appropriate.

4.1 BEAB for BS EN 60335 series, household and similar appliances

BEAB Document 40, Test Parameters for Appliances covered by the BS EN 60335 series of Standards

References: BS EN 60335-1 : 1994 Safety of Household Electrical Appliances
BS EN 50106 : 1997 Routine Tests for Household Appliances

1 Introduction

These tests are intended to reveal a variation during manufacture which could impair the safety of household electrical appliances whose construction is covered by Harmonised Standard BS EN 60335. The tests do not impair the product's properties or reliability and are to be performed on every appliance unless otherwise agreed by BEAB or its authorised representative. Testing is normally carried out on the complete product after assembly with only labelling and packaging being performed after the final safety test. However, the manufacturer may perform the tests at a more appropriate stage in production provided that BEAB agrees that the later manufacturing operations will not affect the results.

If a flexible cord is provided the appliance must be tested with the cord fitted. If an appliance fails any of the tests below it must be subjected to all the tests following repair and/or adjustment. The tests specified below are similar to those specified in Harmonised Standard BS EN 50106 but in some cases BEAB recommends different test values in order to ensure that appropriate safety levels are maintained. The specified tests are the minimum tests considered necessary to cover essential safety aspects. It is the responsibility of the manufacturer to decide if additional routine tests are necessary (see Part 1. Paragraph 8). Following completion of the functional tests the appliance shall be subjected to the electrical safety tests with the appliance switched 'ON'.

2 Functional Tests and Load Deviation

All appliances shall be subjected to a function test to verify correct operation. Any abnormal or out-of-limits results shall be fully investigated in order to ensure safety will not be affected. The functioning of a component is checked by inspection, or an appropriate test if a malfunction could result in a hazard. Verifying the direction of rotation of motors or the appropriate operation of switches and controls are examples of checks which may be necessary.

BEAB recommends that the power input of every appliance should be measured to ensure compliance with the appropriate Standard to which the appliance has been approved.

3 Earth Continuity Test

For Class I appliances, a current of at least 10 amps *(BEAB recommends 25 amps)* derived from a source having a no-load voltage not exceeding 12 volts, is passed between each of the accessible earthed metal parts and either of the following points:-

- The earthing pin or earthing contact of the supply cord plug
- The earthing terminal of appliances intended to be connected to fixed wiring

- The earthing contact of the appliance inlet

The factory-applied limit for appliances with a supply cord is not to exceed 0.2 Ω ((0.1 + R) Ω, *where R = the resistance of the supply cord*)

For all other appliances the resistance shall not exceed 0.1 Ω.

The test is only carried out for the duration necessary for the measurement to be made; this is particularly important where the nominal cross-sectional area of the cables is less than 0.75 mm^2.

BEAB Recommendation

The test of the Standard (carried out with a current of 25 amps) was designed to determine two things:

(a) That the earthing path has a sufficiently low resistance

(b) That the fuse in the appliance plug or the supply circuit will blow before a weak point in the earthing path (e.g. a path made buy a single-strand connection).

The test specified in BS EN 50106 will check (a) but not (b). It is BEAB's view that the hazard presented by a lack of integrity in the earthing path should be detected in routine production tests, and that the test of the Safety Standard should be applied. There is no evidence that using a current of 25 amps damages the appliance, but if a manufacturer has an appliance where this is a concern, then a lower value can be considered for that specific case. It is BEAB's experience that this is very seldom necessary.

5 Electric Strength Tests

The insulation of the appliance is subjected to a voltage of substantially sinusoidal waveform having a frequency of 50 Hz or 60 Hz for 1 s. No breakdown or flash-over shall occur during the test. The value of the test voltage (rms) and the points of application are shown in Table IVa below.

The insulation resistance test and dielectric strength test may be combined into a single test by using an electric strength test set which incorporates a current sensing device which normally trips when the current exceeds 5 mA. Tripping the sensing device shall activate an audible or visual indication of breakdown of the insulation. The high voltage transformer shall be capable of maintaining the specified voltage until the tripping current flows.

BEAB recommend that the leakage (touch) current limit should be set to the equivalent minimum insulation resistance requirement for the product as given in the Standard (e.g. Test voltage of 3750 V rms and insulation resistance of 7 Megohms = leakage (touch) current limit of 0.54 mA rms).

Table IVa - Electric Strength Tests

Points of Application	Test Voltage V Class I Appliances	Class II Appliances	Class III Appliances
1 Between live parts and accessible metal parts separated from live parts by			
• Basic insulation only	1000 *(1250)*		
• Double or reinforced insulation	2500	2500 *(3750)*	400 *(500)*
2 Between live parts and metal parts separated from live parts by basic insulation only		1000 *(1250)*	
3 Between inaccessible metal parts and the body		*(2500)*	

Note: A d.c. test voltage may be used instead of a.c., the values of the d.c. test voltages shall then be 1.5 times those shown in the Table.

(BEAB recommends using the values shown in italics)

6 Additional Requirements for Microwave Ovens covered by the BS EN 60335-2-25 Safety Standard

Electric Strength Test – the Electric Strength Test Equipment may have the current sensing device set up to 100 mA.

Check that the warnings concerning microwave energy specified in BS EN 60335-2-25 are marked on the relevant covers.

Check that the appliance is provided with the correct instructions for that particular appliance.

Check the operation of the door interlock system to ensure that microwave generation ceases when the door is opened.

The microwave oven is operated at rated voltage and with microwave power set to maximum. The energy flux density of microwave leakage is measured at any point approximately 5cm from the external surface of the appliance. An appropriate load may be used. The measuring instrument is moved over the surface of the oven to locate the points of maximum leakage, particular attention being given to the door and its seals.

The microwave leakage shall not exceed 50 W/m^2.

BEAB recommends:

(a) that the appliance is operated at the upper end of the voltage range,

(b) that the test load should be 1000 g +/-5 g of potable water or equivalent,

(c) that the microwave leakage monitor be regularly calibrated (recommended frequency = monthly) against a standard leakage source which itself is calibrated annually. Records of all checks are to be made and these shall be available for inspection.

4.2 BEAB for BS EN 60065, audio, video and similar appliances

BEAB Document 40, Test Parameters for Electronic Equipment covered by BS EN 60065

References: BS EN 60065 : 1998 Audio, Video and Similar Electronic Apparatus – Safety Requirements

1 Introduction

These tests are intended to reveal a variation during manufacture which could impair the safety of Household Electronic Apparatus, connected to the mains supply for indoors use, the construction of which apparatus is covered by Harmonised Standard BS EN 60065. The tests do not impair the properties or reliability of the apparatus.

2 Function Test

Functional or performance tests are to be carried out as considered necessary by the manufacturer prior to the final electrical safety test being performed.

3 General Considerations

The electrical safety tests listed are normally to be carried out, as applicable, 100% at the final stage of manufacture prior to packaging. However, the manufacturer may perform the tests at an appropriate stage during production providing that the BEAB inspector agrees that the later manufacturing operations will not affect the results.

4 Test Method

For all tests, with the exception of the Earth Continuity test on Class I equipment, any mains switch must be in the ON position.

Due account should be taken of the charging time for the total capacitance of the product under test particularly when using d.c. test voltages.

Test failures are to be indicated by visual or audible means.

5 Earth Continuity Test

Products of Class I construction that have accessible metal parts which may become live in the event of an insulation fault shall be tested as follows:

A current of 10 A (BEAB recommendation 25 A See Note Part 2a Para 3), derived from an a.c. source having a no-load voltage not exceeding 12V, is passed between those metal parts and the earth conductor of the mains supply cord for 3 seconds. The resistance shall not exceed $0.1+R\ \Omega$, where R is the resistance of the mains supply cord.

6 Insulation Resistance

The insulation resistance is to be measured on all products between both poles of the mains supply cord or terminals connected together and accessible metal parts. The test is applied using the voltage and limits shown in Table 4.

7 Electric Strength Test

This test is conducted between live parts, with both poles of the mains supply cord or terminals connected together, and accessible metal parts using a voltage of substantially sine wave form, having a frequency of 50/60 Hz and the value as shown in Table IVb or a d.c. voltage equivalent to 1.414 times the rms voltage, applied for 6 seconds or for 3 seconds if the test equipment incorporates an automatic indication of insulation breakdown and has to be manually reset by the operator.

Table IVb - Test Requirements for Electric Strength Tests

Type of Test	Test Value Applied	Factory Limit Applied
a) Insulation Resistance		Minimum of
Class I Products	500 V d.c.	2 M Ω.
Class II Products	500 V d.c.	4 M Ω.
b) Electric Strength		
Basic Insulation	1500 V rms	No flash-over
Supplementary Insulation	2500 *(3000)* V rms	or breakdown
Reinforced Insulation	2500 *(3000)* V rms	shall occur

Notes:

(1) Class II areas in Class I products are to be subjected to the Electric Strength Test Value for Supplementary/Reinforced Insulation.

(2) If the Electric Strength test equipment has an automatic indication of insulation breakdown which must be manually reset, the trip current setting must not normally exceed *6 mA rms*.

(3) BEAB recommends the use of the test values shown above in italics.

8 Combined Test

The Insulation Resistance and Electric Strength tests may be combined using an electric strength test set incorporating a sensitive current trip set to show a fail result when it exceeds 0.75 mA at 1500 V rms (Class I) or 0.75 mA at 3000 V rms (Class II).

9 Specific Requirements for Television Room Aerials

Television room aerials approved to BS 5373 shall be subjected to safety tests covering Insulation Resistance and Electric Strength. For further details of these tests contact BEAB.

4.3 BEAB for BS EN 60950, information technology and similar appliances

BEAB Document 40, Test Parameters for Information Technology Equipment covered by BS EN 60950

References: BS EN 60950 : 1992 Safety of Information Technology Equipment, Including Electrical Business Equipment

1 Introduction

These tests are intended to reveal a variation during manufacture which could impair the safety of mains or battery powered IT equipment for home or business use, the construction of which equipment complies with Harmonised Standard BS EN 60950. The tests do not impair the properties or reliability of the apparatus.

2 Function Tests

The equipment shall be subjected to functional or performance tests as considered necessary by the manufacturer prior to the final electrical safety test being performed.

3 General Considerations

The electrical safety tests listed are to be carried out, as applicable, 100% at the final stage of manufacture prior to packaging unless otherwise agreed. However, the manufacturer may perform the tests at an appropriate stage during production provided that the BEAB inspector agrees that the later manufacturing operations will not affect the results.

In addition to the requirements shown in Part 1, Certificates of Conformity for materials and components must confirm that the following applies:

a) High voltage components comply with BS EN 60950 Clause 4.4 or equivalent.

b) Capacitors that bridge insulation comply with BS EN 60950 Clauses 1.5.6 / 1.6.4 and IEC 60384 Part 14 (latest edition).

Plastic parts used in construction and printed circuit boards comply with the current Flame Retardancy Classification (FRC), such as granted by Underwriters Laboratories Inc.

4 Test Method

For all tests, with the exception of the Earth Continuity test on Class I equipment, any mains switch must be in the **ON** position.

Due account should be taken of the charging time for the total capacitance of the product under test particularly when using d.c. test voltages. **Note**: The Insulation Resistance test of BS EN 60065 is replaced by an Earth Leakage test for BS EN 60950 products.

Test failures are to be indicated by visual or audible means.

5 Earth Continuity Test

Products of Class I construction that have accessible metal parts which may become live in the event of an insulation fault shall be tested as follows:

A current of 10 A (BEAB recommendation 25 A See Note Part 2a Para 3), derived from an a.c. source having a no-load voltage not exceeding 12 V, is passed between those metal parts and the earth conductor of the mains supply cord for 3 seconds. The resistance shall not exceed 0.1+R Ω where R is the resistance of the supply cord.

6 Earth Leakage Test

This test is carried out at the rated mains input voltage and the maximum leakage shall not exceed the following:

Class I products – 3.5 mA (0.75 mA for hand-held appliances)

Class II products – 0.25 mA

7 Electric Strength Test

This test is conducted between live parts, with both poles of the mains supply cord or terminals connected together and accessible metal parts using a voltage of substantially sine wave form, having a frequency of 50/60 Hz and the value shown in Table IVc, or a d.c. voltage equal to 1.414 times the rms voltage, applied for 6 seconds or for 3 seconds if the test equipment incorporates an automatic indication of insulation breakdown and has to be manually reset by the operator.

Table IVc - Test Requirements for Electric Strength Tests

Type of Test	Test Value Applied	Factory Limit Applied
a) Earth Leakage		
Class I: Hand-Held Appliances	Rated	0.75 mA
Other Class I Appliances	Input	3.5 mA
Class II Appliances		0.25 mA
b) Electric Strength		
Operational Insulation	1500 V	No flash-over
Basic Insulation	1500 V	or breakdown
Supplementary Insulation	1500 V	shall occur
Reinforced Insulation	3000 V	

Notes to Table IVc:

(1) Insulation breakdown is considered to have occurred when the insulation does not restrict the uncontrolled flow of current.

(2) When testing equipment incorporating solid state components that might be damaged by the secondary effect of the testing, the test may be conducted without the components being electrically connected providing that the wiring and terminal spacings are maintained.

8 Combined Test

The Earth Leakage and Electric Strength tests may be combined using an electric strength test set incorporating a sensitive current trip set to show a fail result when it exceeds 20 mA for Class I and 3 mA at 3 kV for Class II products.

4.4 BABT

BABT Document 440, Electrical and Acoustic Safety Test

TABLE IVd - BABT ELECTRICAL AND ACOUSTIC SAFETY TESTS

No	Test Connection Points	Test Title	Test Condition	Test Limits	Notes
1	Connection point 1: Main protective earth connection within equipment. Connection point 2: Other user accessible parts of equipment which have been connected to protective earth for safety reasons (and are hence protectively earthed).	Earth continuity	Max test voltage: 12 V a.c. or d.c. Min test current: 1.5 times current rating of the primary fuse. Max test current 25 A	Measured resistance to be 0.1 ohm or less	1, 2, 3, 12-15
2	Connection point 1: Phase and neutral conductors shorted together. Connection point 2: Protective earth connection.	Electric strength for basic insulation	Test voltage: 1500 V a.c. or 2121 V d.c. Test time: 2 secs min, 6 secs max	No breakdown	4, 7, 8, 9, 10, 12-15
3	Connection point 1: NTP connectors shorted together. Connection point 2: Conductive parts separated from the NTP by basic or supplementary insulation, shorted together.	Electric strength for basic and supplementary insulation	Test voltage: 1500 V a.c. or 2121 V d.c. Test time: 2 secs min 6 secs max	No breakdown	5, 7, 8, 10, 12-15
4	Connection point 1: Phase and neutral conductors shorted together. Connection point 2: Unearthed user accessible conductive parts or unearthed SELV outputs of a power supply shorted together.	Electric strength for reinforced insulation	Test voltage: 1500 V a.c. or 2121 V d.c. Test time: 2 secs min 6 secs max	No breakdown	6, 7, 8, 9, 10, 12-15
5	Connection point 1: Phase and neutral conductors shorted together. Connection point 2: NTP connectors which are not protectively earthed, shorted together.	Electric strength for reinforced insulation	Test voltage: 3000 V a.c. or 4242 V d.c. Test time: 2 secs min 6 secs max	No breakdown	6, 7, 8, 9, 10, 12-15

6	Connection point 1: NTP connectors shorted together. Connection point 2: Conductive parts, protective earth and auxiliary ports complying with the limits of SELV shorted together.	Separation between interface I_a and user accessible ports	Test voltage: 1000 V a.c. or 1414 V d.c. Test time: 2 secs min 6 secs max	No breakdown	10, 11, 12-15
7	Where acoustic shock protection relies on specific components and possibly their correct orientation then the integrity of these circuits must be verified.	Acoustic shock	BS 6450:Part 2 : 1983: Clause 6.2.10 or TBR8: Annex C or BS 6317 : 1982: Clause 13.9 or 85/013: Issue 4: Clause 5.2.9	+224 dBPa	14

Electrical Safety Tests: Notes

(1) The test as described is for Class I equipment, i.e. equipment which relies on a protective earth connection for providing safety. (This test is in accordance with BS EN 60950, Clause 2.5.11).

(2) As an alternative location for connection point 1, for equipment incorporating a mains supply cord the supply earth connection (normally the earth pin of the mains plug) of the cord shall be used. In this case the measured resistance shall be not greater than (0.1 + R) ohm, where R is the resistance of the earth lead within the mains supply cord.

(3) On equipment where the protective earth connection to a sub-assembly or to a separate unit is by means of one core of a multicore cable which also supplies mains power to that sub-assembly or unit, the resistance of the protective earthing conductor in that cable shall not be included in the resistance measurement (as is the case with the mains cord resistance, see note 2). Where the cable is protected by a suitably rated protective device (which takes into account the impedance of the cable), the minimum test current may be reduced to 1.5 times the rating of this protective device.

(4) The test as described is for Class I equipment. Normally the protective earth connection is the earth pin of the mains plug.

(5) This test is applicable to Network Terminating Points (NTPs) connecting to either analogue or digital networks, but see note 11.

(6) Where the user accessible conductive part, or the NTP connection, is isolated from primary circuits by reinforced or double insulation, but is either:

a) connected to protective earth for functional reasons or

b) separated from protective earth by less than supplementary insulation,

then it may not be possible to conduct this test without overstressing basic insulation (which is only designed to withstand 1500 V a.c. - for further explanation see the relevant notes to BS EN 60950, clause 5.3.2). In this case, the individual components providing the reinforced or double insulation shall be tested in accordance with test No 5, but the finished equipment may be tested in accordance with test No 2. When testing such finished equipment, any user accessible conductive parts which are not connected to earth in the equipment shall be connected to protective earth when performing test No 2.

(7) The test voltage specified may be increased at the discretion of the manufacturer. However, BABT do not require any higher test voltages.

(8) The pass/fail general criteria for electric strength tests is that no breakdown shall occur. In practice, the trip current on the test instrument will need to be set so that it does not trip when subject to the normal leakage (touch) current (predominant for a.c. testing) or insulation resistance current for the equipment and test voltage concerned. Trip current levels should be set to a minimum practical level.

(9) For an item of equipment supplied with a UK 13 A mains plug having a leakage of 3.5 mA or less, the following maximum trip currents are acceptable in accordance with BS EN 60950. For hand-held Class II equipment lower trip currents are appropriate. See BS EN 60950: clause 5.2.2.

Test Voltage	1500 V a.c.	3000 V a.c.	2121 V d.c.	4242 V d.c.
Trip Current	42 mA	84 mA	1.5 mA	3.0 mA

(10) The two second minimum for the test duration shall apply where operator judgement against a time standard is used. Where the timing is carried out automatically, the minimum test time may be reduced to one second (the one second minimum is in accordance with the note regarding production tests in BS EN 60950 clause 5.3.2). The six second maximum is advisory. However, BABT deprecate longer production test times and repeated electrical strength testing as they can cause damage to insulation.

(11) This test is applied only to Digital Terminating Equipment as an option to test No 3.

(12) Where appropriate, the sequence of safety tests shall be earth continuity followed by electric strength.

(13) Where the integrity of each test may be demonstrated, it is possible to combine certain of these tests.

(14) Where tests on sub-assemblies have been conducted under BABT surveillance, it is not necessary to re-test the complete assembly if overall compliance with the appropriate tests is demonstrated to BABT and if final assembly arrangements do not affect the integrity of the sub-assembly tests.

(15) When the above tests are to be applied to Power Supply Units the low voltage outputs are considered to be the NTP.

APPENDIX V **MODEL FORMS for In-service Inspection and Testing**

Form Va **Equipment Register**

Equipment register					
Register No	Location	Equipment description	Serial No	Frequency	
				Formal visual inspection	Combined inspection and testing

Form Vb

Equipment Formal Visual and Combined Inspection and Test Record

Equipment Inspection and Test Record				1 Register No	
2 Description of equipment	3 Construction Class	4 Equipment type	5 Location and particular requirements of location	6 Frequency of Formal visual inspection .... m	Combined inspection and testing .... m
7 Make .... *Model .... Serial No	8 Voltage .... V Rating .... A Fuse Rating .... A	9 *Date of purchase		10 *Guarantee	

Inspection							Test							
11 Date	12 Environment /use	13 Dis-connect	14 Socket	15 Plug	16 Flex	17 Body	18 Continuity		19 Insulation		20 Functional check	21 Comments/other tests	22 OK to use	23 Signature
							Ω	✓	MΩ	✓				

Note: (✓) Indicates pass (x) Indicates fail (N/A) Not applicable (N/C) Not checked
*To be completed by client

Notes on Inspection and Test Record

Notes:

1. Register No - this is an individual number taken from the equipment register, for this particular item of equipment.
2. Description of equipment, e.g. lawnmower, computer.
3. Construction Class - Class 0, 0I, I, II, III. Note that only Classes I and II may be used without special precautions being taken.
4. Equipment types - portable, movable, hand-held, stationary, fixed, built-in.
6. Frequency of inspection - generally as suggested in Table 1 of the Code of Practice.
   Inspection - inspection items 11-17 and 20 to 23 will be completed if an inspection is being carried out.
   Inspection and Test - when testing is carried out, this must be preceded by the inspection items.
11. Date of 'Inspection' or 'Inspection and testing'.
12. Environment and use. It must be confirmed that the equipment is suitable for use in the particular environment and is suitable for the use to which it is being put.
13. Authority is required from the user to disconnect equipment such as computers and telecom equipment where unauthorised disconnection could result in loss of data.
    Authority must also be obtained if such equipment is to be subjected to the insulation resistance and electric strength tests.
14. Socket/flex outlet - the socket or flex outlet must be inspected for damage including overheating.
    If there are signs of overheating of the plug or socket, the socket connections must be checked as well as the plug. This work should only be carried out by an electrician.

15, 16, 17 The inspection required is described in Section 14 of the Code of Practice for In-Service Inspection and Testing of Electrical Equipment published by the IEE.

18, 19 Tests - these are described in Section 15 of the Code of Practice for In-Service Inspection and Testing of Electrical Equipment. They must always be preceded by the Inspection items 11-17. The instrument reading is to be recorded and ✓ entered if the test results are satisfactory.

20-23 These columns are to be completed for inspection only as well as inspection and testing.

20. Functional check - a check is made that the equipment works properly.
21. Comments/other tests - to identify failure more clearly, and to indicate other tests carried out, e.g. touch current measurement.
22. OK to use - 'YES' must be inserted if the appliance is satisfactory for use, 'NO' if it is not.

Form Vc **Equipment labels**

A. LOGO

PASS

Date of check ____________________

Initials ____________________

Appliance No ____________________

Next test before ____________________

SAFETY CHECK

Form Vd **Repair Register**

Repair register							
Register No	Customer	Description	Serial No	Repairer	Suitable for return to use		
					✓	Signature	Date

(✓) Indicates satisfactory (x) Indicates unsatisfactory

Form Ve **Faulty equipment register**

Register of faulty equipment				
Date	Register No	Equipment fault	Location	Actioned

APPENDIX VI

RESISTANCE OF FLEXIBLE CABLES

TABLE VI Nominal resistances of appliance supply cable protective conductors (figures are for cables to BS 6500/BS 6360)

Nominal conductor csa (mm^2)	Nominal conductor resistance at 20 °C (mΩ/m)	Length (m)	Resistance at 20 °C (mΩ)	Max current-carrying capacity (A)	Max dia of individual wire in conductor (mm)	Approx No of wires in conductor
0.5	39	1	39	3	0.21	16
		1.5	58.5			
		2	78			
		2.5	97.5			
		3	117			
		4	156			
		5	195			
0.75	26	1	26	6	0.21	24
		1.5	39			
		2	52			
		2.5	65			
		3	78			
		4	104			
		5	130			
1.0	19.5	1	19.5	10	0.21	32
		1.5	29.3			
		2	39			
		2.5	48.8			
		3	58.5			
		4	78			
		5	97.5			
1.25	15.6	1	15.6	13	0.21	40
		1.5	23.4			
		2	31.2			
		2.5	39			
		3	46.8			
		4	62.4			
		5	78			
1.5	13.3	1	13.3	15	0.26	30
		1.5	20			
		2	26.6			
		2.5	33.3			
		3	39.9			
		4	53.2			
		5	66.5			
2.5	8	1	8	20	0.26	50
		1.5	12			
		2	16			
		2.5	20			
		3	24			
		4	32			
		5	40			
4	5	1	5	25	0.31	53
		1.5	7.5			
		2	10			
		2.5	12.5			
		3	15			
		4	20			
		5	25			

The above Table gives figures for the nominal resistance of the protective conductor per metre length and for various lengths of cable that may be fitted as supply leads to appliances. Once an Earth Bond Test has been performed the approximate resistance of the protective conductor can be found and deducted from the test result to give a more realistic figure for the resistance of the earth bonding of the appliance.

Note: 1000 mΩ = 1 Ω

INDEX

Code of Practice for In-service Inspection and Testing of Electrical Equipment

2nd edition 2001, ISBN 0 85296 776 4

These ERRATA have been corrected in this reprint.
This page is provided to help trainers and lecturers.

ERRATA (Feb 2002)

Contents, Pages 3 to 5, Amend page numbers as follows:

Add 1 to all page numbers, except page 10 which should be 12.

Page 27, Add to heading of 10.5 as follows:

10.5 Earth leakage measurement
(Protective conductor and touch current measurement)

Page 43, Amend 3rd paragraph as follows:

Supply cords must should not be extended …

Page 44, Amend Note (2) to Table 2 as follows:

… appliances with relatively high earth leakage protective conductor current, otherwise …

Page 44, Amend 15.6 as follows:

15.6 Protective conductor/ touch current measurement

Protective conductor/ touch current measurement is an alternative to the in-service insulation test for use if the insulation resistance test either cannot be carried out or gives suspect test results. The Touch current is measured from live parts to earth for Class I equipment, or from live parts to accessible surfaces of Class II equipment.

The touch current is to be measured within 5 seconds after the application of the test voltage and must not exceed the values in Table 3.

For practical purposes the test voltage is the supply voltage.

TABLE 3 - MEASURED PROTECTIVE CONDUCTOR/ TOUCH CURRENT

Appliance Class	Maximum Touch Current note (1)

Page 46, Amend 15.10 as follows:

… Class I appliances with the addition of a polarity check. Any such extension leads …

Page 63, Amend Table IVb, row b) as follows:

Reinforced Insulation 2500 (3000) V rms shall occur

Index, Pages 78 to 81, Amend page numbers as follows:

H	Hotels	Table 1	2	20
L	Legal requirements	3	11	12
T	Testing earth continuity	15.4	35	43
T	Touch Current	2	9	11

Errata published by IEE, 21 February 2002